GIACOMO COCCONI

Numerical Study on the Propagation of Turbulent Fronts in Dilute Polymer Solutions

SCHRIFTENREIHE DES INSTITUTS FÜR STRÖMUNGSMECHANIK
KARLSRUHER INSTITUT FÜR TECHNOLOGIE (KIT)

BAND 5

Eine Übersicht aller bisher in dieser Schriftenreihe erschienenen Bände finden Sie am Ende des Buchs.

Numerical Study on the Propagation of Turbulent Fronts in Dilute Polymer Solutions

BY
GIACOMO COCCONI

KIT Scientific Publishing

Karlsruher Institut für Technologie
Institut für Strömungsmechanik

Numerical Study on the Propagation of Turbulent Fronts
in Dilute Polymer Solutions

Zur Erlangung des akademischen Grades eines Doktor-Ingenieurs
von der KIT-Fakultät für Maschinenbau des
Karlsruher Instituts für Technologie (KIT) genehmigte Dissertation

von M.Sc. Giacomo Cocconi

Tag der mündlichen Prüfung: 12. Dezember 2017
Referenten: Prof. Dr.-Ing. Bettina Frohnapfel, Prof. Dr. Jörg Schumacher

Impressum

KIT Scientific Publishing

Karlsruher Institut für Technologie (KIT)
KIT Scientific Publishing
Straße am Forum 2
D-76131 Karlsruhe

KIT Scientific Publishing is a registered trademark
of Karlsruhe Institute of Technology.
Reprint using the book cover is not allowed.

www.ksp.kit.edu

Print on Demand 2019 – Gedruckt auf FSC-zertifiziertem Papier

ISSN 2199-8868
ISBN 978-3-7315-0829-8
DOI 10.5445/KSP/1000085139

Abstract

The purpose of this work is to expand the current knowledge about the mechanism with which turbulent regions interact with non-turbulent ones. The continuous exchange of mass and momentum between these two regions is a process characterized by some of the smallest and the largest scales on the flow. In such a spectrum of scales, viscous diffusion of enstrophy in the past has been considered to be the drive of turbulent propagation through small scale vortices diffusing into the nearby laminar fluid. Inertial dynamics, on the other hand, are considered to play a role only in the measure in which they are able to increase or decrease the total surface of the interface over which vorticity viscously diffuses into the laminar region. In order to better assess these hypotheses, the present study recurs to numerical simulations of turbulent fronts with zero mean shear. First, a scale analysis has been performed by studying the spectra of the enstrophy budget equation across the interface. Secondly, the effect of an altered scales distribution is investigated. This is achieved without directly affecting the viscous mechanics, thus by using dilute polymer solutions. Spectral analysis reveals not only that the inertial transport of turbulent fluctuations holds a central role in sustaining the interface propagation, but also that viscous diffusion is characterized by two scales: a thickness of the order of the Kolmogorov scale in direction normal to the interface and a larger width of the order of the Taylor length scale. Polymer solutions, simulated by means of the FENE-P model, show to produce interfaces with smoother features and larger scales, more importantly turbulent fronts in dilute polymer solution propagate less than their Newtonian equivalents. Evidence shows a local action of the polymers at the interface via preferential alignment with and enhancement of vortex compression.

Kurzfassung

Ziel dieser Arbeit ist die Erweitung des aktuellen Wissens über den Interaktionsmechanismus zwischen turbulenten und laminaren Regionen. Der kontinuierliche Austausch von Masse und Impuls zwischen den beiden Regionen ist ein Prozess, der sowohl durch die kleinsten als auch durch die größten Skalen gekennzeichnet ist. In einem solchen Skalenspektrum wurde in der Vergangenheit die viskose Diffussion der Enstrophie als Antrieb der turbulenten Ausbreitung kleinskaliger Wirbel in das benachbarte laminare Gebiet betrachtet. Trägkeitseffekte, hingegen, spielen nach dieser Ansicht lediglich bezüglich der Vergrößerung oder Verkleinerung der gesamten Grenzfläche, über welche die Wirbelstärke in den laminaren Bereich diffundiert, eine Rolle. Um diese Hypothesen besser zu bewerten, wurden in der vorliegenden Arbeit numerische Simulationen turbulenter Grenzgebiete ohne mittlere Scherung eingesetzt. Zuerst wurde eine Skalenanalyse durchgeführt, indem die Spektren der Enstrophie-Bilanz-Gleichung aus der gesameten Grenzfläche zwischen turbulenten und nicht turbulenten Regionen untersucht wurden. Als zweites wurde die Auswirkung einer veränderten Skalenverteilung untersucht. Dies wurde durch den Einsatz verdünnter Polymerlösungen erreicht, welche die viskose Mechanik nicht direkt beeinflussen. Die spektrale Analyse zeigt welch zentrale Rolle der Trägheitstransport turbulenter Fluktuationen beim Erhalt der Grenzflächenausbreitung spielt. Des Weiteren wird gezeig, dass die viskose Diffusion durch zwei Skalen charakterisiert ist: einer dicken in der Größenordnung der Kolmogorov Länge und normal zur Grenzfläche und einer breiten, die größer ist als die dicke un welche, der Größenordnung einer Taylor-Längenskala ist. Polymerlösungen, die mit Hilfe des FENE-P-Modells simuliert wurden, zeigten, dass sie Grenzflächen mit glatteren äußerungen und größeren Skalen erzeugen, vor allem turbulente Fronten in verdünnter Polymerlösung breiten sich weniger aus als ihre Newtonschen äquivalente. Der Nachweis zeigt, wie die Polymere an der Grenzfläche durch eine bevorzugte Ausrichtung und eine Verstärkung der Wirbelkompression lokal wirken.

Contents

1 Introduction **1**

1.1 General aspects of turbulent/non-turbulent interfaces 1

1.2 Turbulent flows of dilute polymer solutions 5

1.3 Objectives and procedure 7

2 Propagation of turbulent fronts in Newtonian fluids . . . **9**

2.1 Role of strain and vorticity in turbulence 10

2.2 Scale dynamics of enstrophy in a propagating shearless turbulent front 15

2.2.1 Enstrophy budgets for a planar symmetric turbulent front 18

2.2.2 Spectral Enstrophy Budget 21

3 Dilute polymer solutions models **33**

3.1 Dilute polymer solutions and turbulent flows 33

3.2 The kinetic theory 36

3.3 The FENE model and the Peterlin closure 42

3.3.1 Limitations of FENE-P model 45

3.3.2 Numerical issues: the high Weissenberg number problem 46

4 Simulation of a turbulent front in dilute polymer solutions **49**

4.1 Set-up of the numerical experiment 49

4.2 Modelling oscillating grid turbulence 50

4.2.1 Properties of oscillating grid turbulence 50

4.2.2 Body force model implementation 52

4.3 Validation and flow properties 54

4.4 Concluding Remarks 62

5 Results and discussion **63**
5.1 Eulerian statistics 63
5.1.1 Conditional statistics 68
5.1.2 Strain and vorticity 73
5.2 Lagrangian Statistics 82
5.2.1 Entrainment 83
5.2.2 Extrainment 86

6 Conclusions **91**

A Appendix **95**
A.1 Numerical Implementation 95
A.2 Forcing characterization 97
A.3 Choice of the parameters for the FENE-P model 100
A.4 Interface Detection Methods and conditional averages 103

Bibliography **109**

Conference Contributions and Publications **119**

1 Introduction

1.1 General aspects of turbulent/non-turbulent interfaces

In most real life problems related to the flowing of a fluid, one eventually will have to deal with turbulent flows. Even today, the highly non-linear and chaotic behavior of turbulent flows constitute a formidable challenge to both theoretical and applicative engineering, leaving a number of open issues in disciplines ranging from material processing to astrophysics. A large subclass of turbulent flows is represented by those ones in which turbulent-flow regions and regions in which the flow is laminar coexist. This is the case, for example, of boundary layers, jets, shear flows, plumes and wakes, without forgetting all kind of transitional flows. The boundary separating the laminar and turbulent regions, often referred to as turbulent/non-turbulent interface (or for brevity also TNTI), is a highly convoluted and non-stationary surface. The two regions possess striking differences, among which one of the most relevant is probably the large change in mixing rates: within a turbulent flow quantities as momentum, temperature, passive scalars or reactants concentrations are spread and mixed at much faster rates [19], while in laminar flow mixing is brought about by the several orders of magnitude slower molecular diffusion. Therefore an enhanced spreading of the turbulent region is obviously sought-for in all applications where such increased mixing is desirable and avoided in those where it is not. Examples of the former case are chemical reactors, heat exchangers or turbulators on wings, while in the latter group of applications one can find the drag-reduction, spillage containment, reduction of sediment re-suspension or brown-out, erosion prevention. The passage itself from a turbulent to a non-turbulent region is sharp and abrupt [19], the differences are so striking that in several cases it is possible to discern the boundary between this regions without any particular expedient. This happens, for example, for the transport of passive scalars whose diffusivity is two to three orders of magnitude smaller than the kinematic viscosity of the ambient fluid [19]. In such regime, the turbulent/non-turbulent interface effectively delimits the volume of fluid within the scalar quantity can be transported. Corrsin and Kistler [17] have been the first to study the properties of the turbulent/non-turbulent interface in a free shear layer; in their work they identified the non-turbulent region as the part of the field deprived of vorticity. They theorized that the weak, large-scale irrotational fluctuations that can be found in the laminar region should smoothly match the intense small scale vortical fluctuations of the turbulent one in a thin region called laminar or *viscous superlayer*. This layer is dominated by the viscous forces which are recognized to be the sole mechanism of propagation of vorticity in the irrotational region [8, 19].

The sharp, quasi-discontinuous character of the turbulent/non-turbulent interface historically made it challenging to study it and direct observations of the laminar superlayer have not been possible until recent times. Most of the early studies had to rely on fixed point measurement of the intermittency of some variable (e.g. velocity, temperature, scalar concentrations) in order to estimate the average position of the interface. With relatively recent developments in experimental techniques, the instantaneous measurements of sufficiently resolved regions of the velocity field next to the interface became available. Almost at the same time the increase of the available computational power had permitted to simulate interfacial flows with resolution up to the Kolmogorov scale. As a consequence, new and more refined ways to study the turbulent/non-turbulent interface had been developed: thanks to these new techniques Bisset et al. [7] have first been able to observe the theorized jump of velocity variance and vorticity across the turbulent/non-turbulent interface. It has been observed that vorticity is the quantity that changes the most sharply across the interface, proving itself as the most reliable way to detect the interface in a number of flows [19, 36, 42]. Other properties, like the concentration of passive scalars or temperature, have similar though less steep variations and have been used in order to detect the interface where vorticity was not directly accessible [7, 36, 42, 97].

Once the interface is identified in a snapshot of the field, it is possible to compute a series of statistics conditioned with respect to the distance from the interface. Conditional averaging generates a new reference system for every point over the interface and performs the average over points which are homogeneous in their distance from the interface. This permits to highlight the sharp changes occurring across the interface, where standard averaging methods would smear out such features by averaging together contributes from both turbulent and non/turbulent regions. Such an approach was used in Holzner et al. [43] as well as in Taveira and da Silva 2013 and 2014 [78, 79] to compute budgets of kinetic energy, vorticity and strain rate across the interface. In particular, using this approach Taveira and da Silva 2014 [79] found the existence of a layer with a thickness in the order of the Kolmogorov scale in which the viscous diffusion of vorticity dominates over all other effects. They identified this layer as the viscous superlayer theorized by Corrsin and Kristler [17].

While the viscous superlayer is characterized by some of the smallest scales in the flow [8] the convolution of the interface with its bulges and pockets is apparently driven by the large scale dynamics of the flow [8]. Indeed, it has been observed that the largest convolutions on the surface of the turbulent/non-turbulent interface are the imprint of large-scale vortices beneath it [7] and large scale perturbation in jet outlets are known to produce the phenomena of bifurcating and blooming jets [71], in which entrainment is greatly increased. The large scale bulges at the interface are possibly quite dependent on the large-scale dynamics that sustain the turbulent flow and may differ in wakes, jets, boundary layers, shear flows and mixing layers [8, 11], even though is still object of debate whether this affects the entrainment rate [96]. Nevertheless it is hard to find a single responsible cause for these changes when comparing such different flows: mean shear, for example, was found to enhance entrainment by increasing both the viscous and the inertial contributions to it [98].

Though this might discourage a generalist approach to the subject, there are also commonalities that are shared by all types of interfacial flows. These might be determined by üniversal"characteristics of turbulence or be due to generic properties specific to the turbulent/non-turbulent interface. An example of both cases is probably its fractal-like nature: a constant fractal dimension of the interface has been found in a range of scales spanning from the integral scale down to near the Kolmogorov scale (Sreenivasan et al. 1989 [76], Chauhan et al. 2014 [11]. Such a fractal dimension appears to remain constant among different type of flows and at different Reynolds numbers. These fractal features sensibly increase the effective surface of the interface [76] and its diffusive flux of vorticity. Another general property of turbulent/non-turbulent interfaces is its strong vorticity anisotropy: this is caused by the solenoidality of the curl, which requires that the vorticity vector remains roughly parallel to any surface across which the vorticity's magnitude drops or increases sharply [8]. This means that vorticity lines cannot cross or end in the irrotational region but are tilted and follow the interface contour. Anisotropy, steep gradients and large span of interplaying scales make turbulent interfaces particularly challenging to both turbulence modelling and LES simulations. This is partly due to these steep unresolved jumps in all the properties and partly due to the fact that the hypothesis of local equilibrium doesn't hold in these regions [19]. This hinders the capabilities of accurately predicting turbulent propagation in those cases where the use of such models is the only viable numerical tool (as it is for most of the medium/high Reynolds number applications). For this reason there is a need for a sounder understanding of the physical interactions between large scale dynamics and the small unresolved ones, as well as between the inertial and the viscous ones.

Beside these applicative necessities, the interface is an interesting region for the study of turbulence *per se*. It can be seen as a problem of turbulence transition, where weak perturbations in the laminar region are amplified under increasing shear and finally destabilize acquiring vorticity once in contact with the interface. At the same time, turbulent fluctuation can undergo the opposite process, i.e. re-laminarization. The difference between the rate of this process and the rate at which irrotational fluid transitions into turbulence determines how fast and far the turbulent region can further spread. Summarizing we can say that in turbulent propagation several concurrent processes contribute to the spreading of the turbulent region across a wide range of flow scales. Small (as well as large) scale vorticity diffusion, though fundamental, cannot be used alone to characterize the propagation rates. Similarly, it is not clear how to relate the differences in the large-scale turbulence generation mechanisms to the differences in the observed entrainment. One interesting way to study the balance of the different processes contributing to entrainment dynamics is perhaps to "tweak"the turbulence in such a way to alter such equilibrium for the same flow topology, i.e. forcing mechanism. For example, one may alter the way energy is redistributed from the energy-containing scales towards the dissipative range without significantly change viscous dynamics. Perhaps one of the most effective ways available to alter the turbulent flow of a Newtonian liquid is the addition of small quantities of long chain polymers.

1.2 Turbulent flows of dilute polymer solutions

There are a number of ways to alter the properties of a turbulent flow by adding additives to the flow. In the search of a way of reducing drag in various wall bounded systems, scientist and engineers experimented with a wide range of substances capable of reducing drag in a turbulent flow. The experimented with fibers, clays, bubbles, paper pulp, surfactants cationic and anionic, dust particles, sand suspensions, flocks, algae and biological molecules as certain long chain polysaccharides, flocculating agents, paramagnetic particles in presence of magnetic fields and polymers [6, 41, 92], just to name a few. There is not, up to today, general agreement on how such a range of materials and techniques can achieve drag reduction in a turbulent flow. This is particularly true for dilute polymer solutions, which have proven to be one of the most effective agents with reported drag reductions in turbulent shear flows up to 80% using dilute solutions with just a few parts per millions of polymers [41]. Until the flow remains below a certain Reynolds number no differences are directly observable between a dilute polymer solution and a Newtonian fluid flow of matching viscosity [87]. But when a certain threshold Re number is reached it is possible to observe a general reduction in the intensity of Reynolds stresses and turbulent production [87].

One of the simplest models of a polymer molecule approximates it as a chain of concentrated masses kept together by elastic links. Velocity gradients of scales comparable to the molecular length would act on these concentrated masses by pulling them apart, stretching the polymers and storing the energy taken from the flow in their bonds. When the shear is removed the polymer returns to a coiled state returning the energy in the dissipative range of the flow or, at an even smaller scale, in form of thermal agitation. One of the main difficulties in the study of turbulence in dilute polymer solutions is the scales of the phenomenon: a few polymer molecules every some million molecules of solvent already affects a turbulent flow, and the molecules used have lengths varying from a few tens of nanometres up to hundreds of micrometres [41]. Up to the present day,the scales of the problem combined with the impracticalities in measuring polymer stresses made it impossible to experimentally study the molecular dynamics of polymers in turbulent flows. Also, on the numerical side, the scales and the large number of added degrees of freedom from the polymer molecules make it computationally prohibitive to simulate even small amount of polymer molecules dispersed in a turbulent flow. Hence, the study of polymer dynamics in turbulent flows has to rely on simplified models and numerical simulations. Despite the intrinsic limitations of this approach and the strong assumptions used, the polymer models available today have been able to qualitatively replicate many of the main aspects of turbulent flows of dilute polymer solutions [67].

Most of the current literature on turbulence of dilute polymer solutions focuses on the reduction of drag and hence on wall-bounded flows. In these flows, polymers appear to be active essentially in the near-wall region, where they are particularly effective in suppressing wall-normal velocity fluctuations and increasing the anistropization of the flow [29]. In general, a polymer in the flow is stretched, un-stretched, tilted and transported by the flow according to its time scales, hence the interaction of a polymer chain with the underlying turbulence is theoretically strongly dependent on the history of stretching and orientation of the molecule. Nevertheless, in experiments where polymers are locally injected in a point-wise manner, it has been observed that, by choosing an injection point in the near-wall region, it is possible to achieve maximum reduction of drag similar to flows with homogeneously distributed polymers [33, 34, 63]. Moreover, it is observed that, by reducing the wall-normal velocity fluctuations, the injected polymers remain confined at about the same distance from the wall for several eddy turnover times downstream of the injection point [63], maintaining their action limited to such region. The importance of the near-wall region is highlighted by the diminished drag-reducing potential of polymers in pipe flows with rough walls [91], suggesting that this phenomenon might be strongly correlated to the organized structures present in the near-wall region of hydraulically smooth wall-bounded flows. Hence, even though the interaction between polymer and fluid is theoretically strongly dependent on the extension history, i.e. non-local in time and space, at least some of its macroscopic effects are indeed restricted to limited regions of the flow. This opens the possibility to at least partially de-couple the flow-polymer interaction from the extension history and allows linking it more directly to the local properties of the flow. In the present study of turbulent/non-turbulent interfaces this partial loss of memory of the polymer leaves open the question of whether the polymer affects the interface through the general alteration of turbulence or whether its action is more localized directly in the interfacial region. The scales of the molecule generally used limit any possible direct action of the polymers to the smallest scales of the flow, although indirect effects from small scales on larger ones are possible under certain conditions via triadic interactions [5, 87]. Dilute polymers are hence an ideal candidate to study the importance of small-scale dynamics at the turbulent/non-turbulent interface. Indeed, they permit the creation of flows with the same large-scale energy injection of the Newtonian case, but with a dissipative/diffusive range shifted towards larger scales. By analysing where and how the polymers affect the flow, it is possible to understand how much the interface is influenced by the local small-scale dynamics and how much it is influenced by the bulk of the turbulence sustaining it. Finally, observing the behaviour of the polymers at the interface will give further insight on the mechanisms with which they suppress turbulence.

1.3 Objectives and procedure

This thesis aims to study the dynamics with which the irrotational fluid nearby a turbulent front transitions to a turbulent state and becomes entrained in the mass of the turbulent flow. As previously observed, this process is characterized by concurrent inertial and viscous dynamics over the full range of scales of the flow. In order to discern mechanisms which are most relevant to the entrainment, a comparison is made between how it unfolds in a classical turbulent fluid against the case of a fluid with non-Newtonian behaviour (specifically, a dilute polymer solution). In these fluids normal turbulence is altered, and the changed interplay between scales affects the turbulent/non-turbulent interface giving new insights on this phenomenon. The objective of the thesis can be hence summarized as follows:

- study the scales at which the inertial and viscous propagation dynamics operate,
- study how the turbulent front is affected by the introduction of polymers,
- determine if the observed modifications are due to either prevalently local or non-local effects.
- assess the importance of the viscous against the inertial dynamics in the propagation of a turbulent front.

In order to do so, a series of direct numerical simulations of propagating turbulent fronts with both, a Newtonian fluid and a dilute polymer solution model have been performed. The choice of DNS has been justified by the fact that turbulent/non-turbulent interface is characterized by both, some of the smallest and larges scales of the flow. Contrary to experimental measurements, DNS simulations provide full 3D fields and resolution over the whole range of scales of the interface. Moreover, through dilute polymer models, the DNS approach gives access to the extensional and orientational status of the polymers in every point of the turbulent flow. This allows to directly analyse how polymers interact with turbulence in different regions of the flow. Being the literature on the behaviour of polymer models with turbulent/non-turbulent interfaces quite limited, validation against experiment is nevertheless still required. To this purpose, the numerical results will be compared with data available from a set of experiments performed by the Turbulence Structure Laboratory of Tel Aviv University. For the purpose of this study, the scope has been limited to turbulent fronts propagating in absence of mean shear, as this reduces the number of parameters to take in account and in certain situations permits to use some simplification derived from homogeneous isotropic turbulence and thus greatly simplifying the analysis of the results. Generality is maintained throughout the work by focusing on those aspect of turbulence and turbulent interfaces which have been found to be common to different types of flow.

In the following chapters, first the propagation of shear-less turbulent fronts in Newtonian fluids is discussed, focusing on the role of strain and enstrophy combined with an analysis of the scales dynamics of viscous and inertial processes. In Chapter 3 properties and models of dilute polymer solutions are introduced. In Chapter 4 the simulations of turbulent fronts in dilute polymer solutions will be first introduced. Chapter 5 proceeds with the discussion of the simulations' results and analyses the effect of polymers on the turublent/non-turbulent interface.

2 Propagation of turbulent fronts in Newtonian fluids

One of the most delicate parts when approaching the study of the propagation of turbulent fronts is the definition of when a parcel of fluid can be considered turbulent. It is a non-trivial matter if one considers that still today there is no consensus on a universal definition of what turbulent flow is (for example see Tsinober 2001 [86] for a generous list of definitions used in literature). An example of the difficulties in finding a definition for the concept turbulence is shown in the definition given by Batchelor [4]: "[...] it is a well-known fact that under suitable conditions, which normally amount to a requirement that the kinematic viscosity ν be sufficiently small, some of these motions are such that the velocity at any given time and position in the fluid is not found to be the same when it is measured several times under seemingly identical conditions. In these motions the velocity takes random values which are not determined by the ostensible, or controllable, or, 'macroscopic' data of the flow, although we believe that the average properties of the motion are determined uniquely by the data. Fluctuating motions of this kind are said to be turbulent."This and most of the available definitions identifie turbulence as a global property of the flow but fall short when there is the need to locally determine whether a portion of the fluid is turbulent or not. Probably the most evident characteristic of turbulence is indeed its fluctuating (both in time and space), chaotic dynamical properties. It therefore is obvious, in the search for some robust turbulence marker, to focus on the tensor of derivatives of the velocity field, in particular on its decomposition in its symmetric and antisymmetric parts, i.e. strain and vorticity. Hereafter it will be shown how vorticity is fundamental for the identification of non-turbulent portions of the flow and how, together with the strain, it is fundamental in the propagation of turbulent fronts.

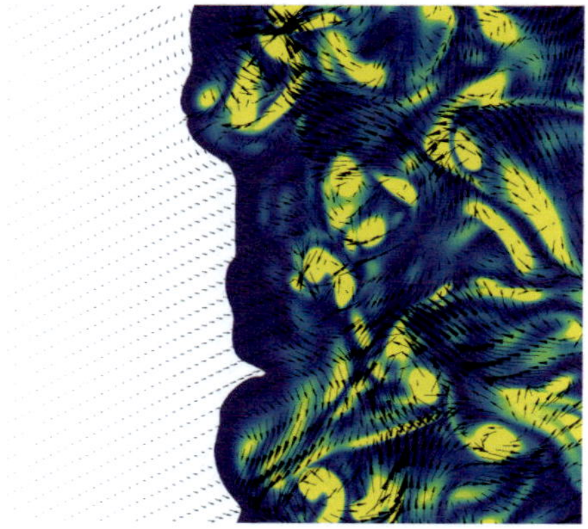

Figure 2.1: Velocity field vectors (arrows) superimposed over isocontour of $\omega_i^2/2$ in a DNS of a propagating turbulent front. The color cut-off level is set at 2% of the average enstrophy in the turbulent bulk of the flow.

2.1 Role of strain and vorticity in turbulence

The strain rate tensor $\mathbf{S} = \frac{1}{2}[\partial u_i/\partial u_i + \partial x_j/\partial x_i]$ is the symmetrical part of the velocity gradient tensor $\mathbf{A} = \nabla\vec{u} = [\partial u_i/\partial x_j]$. Its antisymmetric part is the rotation rate tensor $\mathbf{O} = \frac{1}{2}[\partial u_i/\partial x_j - \partial u_j/\partial x_i]$, where the non-zero elements of $\mathbf{O}$ are the components of the vorticity vector $\vec{\omega} = \nabla \times \vec{u}$. Strain and vorticity play a fundamental role in the amplification and break-down of velocity fluctuations that ultimately lead to turbulence and sustain it. In first place, there is not such a thing as a turbulent flow without vorticity, making it one of the necessary conditions for the existence of turbulence. Another defining element of turbulent flows is the mutual interaction between vorticity and strain. Their nonlinear interactions are indeed responsible for the ability of turbulent flows to draw energy from whatever forcing mechanism is available, redistributing it [38]. Vorticity alone, more than the strain, revealed itself to be one of the most robust ways to discern between regions of turbulent flows from non-turbulent ones [8, 19, 36, 42, 95].

Figure 2.1 shows an example of a TNTI with the velocity field vectors superimposed to the enstrophy field. The white area delimits the region where enstrophy fluctuations drop under 2% of their average value in the core of the turbulent region, while this fluid is to be considered non-turbulent, it still experience some velocity fluctuations. Indeed, the TNTI is highly unsteady and the movement of its offshoots produces these irrotational velocity fluctuations in the non-turbulent region. Such fluctuations decay further away from the interface as $(x - x_I)^{-4}$ [7], where x_I is the local position of the interface, and consequently also viscous dissipation of kinetic energy occurs outside the turbulent region [19]. Despite being unsteady and apparently chaotic these irrotational fluctuations do not carry the characteristic increase in transport properties of turbulent flows. Indeed, when weakly diffusive passive scalars are dispersed in a turbulent flow, the boundaries of the region containing appreciable concentrations of scalar are found to agree very well with the boundary of the rotational region [19, 37]. In the measure to which the turbulent region overlaps over the rotational region of the flow, the study of the propagation of turbulence can be reduced to the study of the mechanics of transport, production, diffusion and dissipation of vorticity fluctuations in the neighbourhood of the interface. The Equation 2.1 for the rate of variation of $\Omega = (\omega_x^2 + \omega_y^2 + \omega_z^2)/2$, also called enstrophy, contains all these contributes for a case without body forces.

$$\frac{\partial \Omega}{\partial t} + u_j \frac{\partial \Omega}{\partial x_j} = \omega_i \omega_j s_{ij} - \nu \frac{\partial \omega_i}{\partial x_j} \frac{\partial \omega_i}{\partial x_j} + \nu \frac{\partial \Omega}{\partial x_j \partial x_j}. \tag{2.1}$$

On the left-hand side of Equation 2.1 one finds the rate of variation of enstrophy and the advection $u_j \frac{\partial \Omega}{\partial x_j}$, responsible for moving about existing enstrophy. The last two terms on the right-hand side are the two viscous contributes $-\nu \frac{\partial \omega_i}{\partial x_j} \frac{\partial \omega_i}{\partial x_j}$ and $\nu \frac{\partial \Omega}{\partial x_j \partial x_j}$, which respectively dissipate and diffuse enstrophy. The average of the first term on the right-hand side, $\omega_i \omega_j s_{ij}$, is found to be always positive in turbulent flows and thus it is usually referred to as enstrophy production [46]. The latter is given by the scalar product $\omega \cdot \mathbf{W}$ between vorticity and the vortex stretching $\mathbf{W} = \vec{\omega}^{\mathsf{T}} \mathbf{S} = \{\omega_j s_{ij}\}$ and is responsible for the coupling between the enstrophy and the total strain rate $s^2 = s_{ij} s_{ij}$. The strain-vorticity interaction is such a central point in turbulence, that Bradshaw [9] includes vortex stretching in its definition: "[t]urbulence is a three-dimensional time-dependent motion in which vortex stretching causes velocity fluctuations to spread to all wavelengths between a minimum determined by viscous forces and a maximum determined by the boundary conditions of the flow". Accordingly three-dimensionality is also a minimum condition for turbulence in 2D flows where this term disappears due to the orthogonality between vorticity and strain, and one cannot talk of turbulence and phenomena as reverse energy cascade due to the fact that vortex coalescence can be observed [38, 46, 86]. The enstrophy production couples the enstrophy equation with the equation for the rate of change of strain, which is given in absence of volume forces by:

$$\frac{1}{2}\frac{\partial s^2}{\partial t^2} + \frac{1}{2}u_j\frac{\partial s^2}{\partial x_j} = -s_{ij}s_{ik}s_{ki} - \frac{1}{4}\omega_i\omega_j s_{ij} + \nu s_{ij}\nabla^2 s_{ij} - s_{ij}\frac{\partial^2 p}{\partial x_i \partial x_j}. \quad (2.2)$$

Here indeed it can be seen how $\omega_i\omega_j s_{ij}$ is a source for the enstrophy equation and a sink for the rate of variation of the strain. Similarly to the enstrophy production the strain has its source term $-s_{ij}s_{ik}s_{ki}$ due to nonlinear self-interactions of strain. Likewise to $\omega_i\omega_j s_{ij}$, strain production is not positive definite and in turbulent flows it is positive only on average, while locally it can contribute to the destruction of strain. The strain equation also contains a dependence to the pressure field through the term $s_{ij}\frac{\partial^2 p}{\partial x_i \partial x_j}$ and a viscous term, which results in contributions equivalent to the ones observed for the enstrophy equation. It has been shown for homogeneous isotropic turbulence that strain and enstrophy equations are mainly driven by the balance between production and viscous terms [58, 81, 86], a balance that for enstrophy has been found to hold locally both in space and time. Those are orders of magnitude higher than their corresponding terms associated with forcing [86] and appear to have universal features among different kinds of flow [58]. Another apparently universal feature is the relative orientation between the local vorticity vector and the strain eigenframe, which appears to follow the same pattern in turbulent flows of different nature and Reynolds numbers [30]. The strain eigenframe can be obtained by decomposing the symmetric strain rate tensor in $\mathbf{S} = \mathbf{Q}\boldsymbol{\Lambda}\mathbf{Q}^{\mathsf{T}}$, where $\mathbf{Q} = \{\vec{\lambda}_1\ \vec{\lambda}_2\ \vec{\lambda}_3\}$ is a tensor with as columns the eigenvectors $\vec{\lambda}_i$ of $\mathbf{S}$ and where $\boldsymbol{\Lambda}$ is the diagonal matrix whose elements Λ_i are the eigenvalues associated to $\vec{\lambda}_i$. The strain eigen-decomposition is closely connected to the enstrophy production and it can be demonstrated that the positiveness of $\omega_i\omega_j s_{ij}$ depends on the signs of the strain rate eigenvalues, since $-3\Lambda_1\Lambda_2\Lambda_3 = -s_{ij}s_{jk}s_{ki} = \frac{3}{4}\omega_i\omega_j s_{ij}$ [86]. The three eigenvalues are identified by means of their relative magnitude: the largest eigenvalue is Λ_1, and it is always found to be positive, while the smallest one, Λ_3, is always negative. Finally, the intermediate eigenvalue Λ_2 can assume both positive and negative values. Therefore, the positiveness of $\omega_i\omega_j s_{ij}$ derives from Λ_2 being on average positive in all turbulent flows. Locally, it is still possible to find negative values of Λ_2 that would lead to locally negative enstrophy production. As can be seen, the definition of the three eigenvalues is purely mathematical, and especially the distinction between Λ_1 and Λ_2 when both are found positive and of similar magnitude should not be considered a physical one, as in a fluctuating field they might switch role rapidly, continuously reverting the axes of the eigenframe. Strain-vorticity orientation and enstrophy production are linked by the relation $\omega_i \omega_j s_{ij} = \omega^2 \Lambda_i \cos^2(\vec{\omega} \cdot \vec{\lambda}_i)$ which means that the orientation between strain eigenframe and vorticity can affect the enstrophy production in a stronger way than the strain and vorticity magnitude alone [86].

The orientations between vorticity and strain eigenvectors are another property that appears to be constant among different turbulent flows [30]. Specifically vorticity is found on average to be preferably aligned with $\vec{\lambda}_2$, weakly aligned with $\vec{\lambda}_1$ and orthogonal to $\vec{\lambda}_3$ [86]. Lüthi et al. [58] showed how these preferential orientations, in particular the fact that $\cos(\vec{\omega} \cdot \vec{\lambda}_i) \approx 1$, are ultimately dependent on the viscous diffusion and destruction of vorticity. The alignment $\cos(\vec{\omega} \cdot \vec{\lambda}_i)$ is indeed more likely to be stronger in those regions of the flow where $\nu\omega\nabla^2\omega$ can be expected to be stronger [58] and viscosity is known to limit the otherwise unbounded growth of the vortex stretching $\omega_i s_{ij}$ [86]. Alignments become particularly important in those regions where the topology of the flow imposes particular orientations between vorticity and strain directly affecting a number of flow properties. Tordella and Iovieno 2011 [83] found that different levels of turbulence among regions of the same flow, and the inhomogeneity originating from it, are enough to re-organize the moments of the velocity derivatives in preferential directions. In their simulations, an initially homogeneous isotropic turbulent flow is manipulated in such a way to reduce the intensity of turbulent fluctuations in half of the domain. What has been observed in this shearless mixing flow is an increase in the mixing region of anisotropy of the velocity derivatives, reduction of compression of fluid filaments parallel to the mixing layer and increase of compression of those orthogonal to it. Regions of different turbulent kinetic energy also differ in their enstrophy content, so this can be another example of the tendency of the vortical lines to tilt in presence of strong enstrophy gradients and to affect turbulent fluctuations accordingly. Such re-organization imposed by inhomogeneities in the flow is most extreme for the case of the TNTI and it will be shown throughout this thesis how this plays a major role in the propagation of the turbulent front.

The relation between viscosity and large scale inhomogeneity had led to the definition of two scalings for the TNTI. Corrsin and Kistler estimated the thickness δ_v of the viscous dominated region of the interface (the viscous super layer) to be in the order of the Kolmogorov scale η. Such scaling was confirmed in several instances both in shearless (using oscillating grids experiments [43]) and sheared unbounded flows (planar jets [78, 98]). The evidence had been found in the position of the peaks of conditional statistics of vorticity and the peak of viscous diffusion close to the interface which all have sizes in the order of few η. It is also evident that the complex large-scale features of the interface are relevant to the interface. The layer mostly characterized by these large scales is the one da Silva et al. [19] refers to as turbulent sublayer which is the "region where the major exchanges between the irrotational fluid and the fully turbulent core occur" [8]. This layer is roughly identified with the region of rapid vorticity magnitude growth [19] and it has been found to scale well with the Taylor microscale λ. The two scalings have been also associated to different mechanics of turbulent propagation: one obviously is the small-scale entrainment due to viscous diffusion of vorticity also called nibbling. Beside this, a large-scale entrainment process takes place when pockets of irrotational fluid are surrounded by large scale structure and advected inside the turbulent bulk of the flow before acquiring vorticity. This process is usually called engulfment [62] and its relevance on the global entrainment rate is still debated [8,96]. The ongoing discussion extends also to the scaling of the TNTI: recently Borrell and Jimenez 2016, while agreeing on a η scaling for the viscous sublayer, demonstrated that in flows characterized by strong shear the observed peaks of vorticity and η scaling can be the product of statistical artifacts [8]. These scaling are evinced from the topology of different conditional averages at the interface, in this way they depend on how conditional sampling is performed and on how the interface is identified in first place. Moreover, the information about the relevant scales at the interface is in this way limited to the scales perpendicular to it. Another approach is to investigate relevant scale of the phenomena leading to the entrainment through their spectral content as it is going to be shown in the following paragraphs.

2.2 Scale dynamics of enstrophy in a propagating shearless turbulent front

As previously noted, vorticity is one of the most evident markers of turbulence and, therefore, it helps in understanding how turbulent fluctuations are generated, amplified, transported and destroyed in the flow. Equation 2.1 gives a measure of exactly this and for the rest of the chapter the turbulent front will be analysed in terms of its enstrophy budget. Focus will be given to the front of a turbulent flow without mean shear as a simplified case representative of many flows, where the turbulence is generated (and possibly sustained) away from the interface. Such case also allows to reduce the number of parameters in the study, as different levels of shear intensity do not need to be accounted for. The analysis is moreover simplified by using one of the most studied and basic classes of turbulent flows: homogeneous isotropic turbulence. Though practically non-realizable, homogeneous isotropic turbulence (HIT) represents an academical approximation of regions of actual turbulent flows, but with independence from boundary conditions and other external influences [38,86]. It is a greatly simplified system in which statistics are invariant with respect to both, translation and rotation, making it a statistically 1D problem. In numerical simulations, HIT can be generated as the natural evolution of the Navier-Stokes equation, starting from a random velocity field or the result of an active stirring, usually via a body force. It has been extensively investigated in literature, because it allows to study fundamental properties of turbulence in a simplified framework with a limited parameter space (essentially reducible to its Re number). It is interesting to study how this prototypical kind of turbulence propagates into quiescent fluid and with this purpose a set of direct numerical simulations have been performed in which TNTI have been added to an initially homogeneous isotropic turbulent flow.

A selected part of the results discussed in this section are published in Cimarelli, Cocconi, Frohnapfel and De Angelis 2015 [14]. In this numerical experiment, the periodic computational domain is first "filled"with homogeneous isotropic turbulence by a stochastic body forcing f_i with a Gaussian distribution centred on the wave number $|\mathbf{k}| = 5$ with variance $\sigma = 0.6$. The dimensionless Navier-Stokes equations:

$$\frac{\partial u_i}{\partial t} + u_j \frac{\partial u_i}{\partial x_j} = -\frac{\partial p}{\partial x_i} + \frac{1}{Re} \nabla^2 u_i + f_i \tag{2.3}$$

have been numerically integrated via a pseudo-spectral solver according to the scheme illustrated in Appendix A.1, $1024 \times 512 \times 512$ Fourier modes have been used for the space discretization of a tri-periodic domain of size $L_x \times L_y \times L_z = 4\pi \times 2\pi \times 2\pi$. For the time discretization, a time step of $\Delta t = 5 \cdot 10^{-5}$ has been used. Initial velocity fields have been selected from a statistically stationary state at $Re_{\lambda_0} = u'\lambda_0/\nu = 120$, where u' is the root mean square of the velocity fluctuations and the initial Taylor length scale has been defined as $\lambda_0 = \sqrt{15u'^2/(2s_{ij}s_{ij})}$. Further details on the simulation parameters are given in Table 2.1. After an initial field has been selected, its velocity fluctuations are artificially damped to zero in half of the domain by multiplying the velocity in every point of the field by a function $p(x) \in [0, 1]$ in such a way as to generate two turbulent/non-turbulent interfaces. Unresolved discontinuities are avoided using a smooth damping function for the velocity fluctuations preventing the appearance of numerical artifacts as Gibbs phenomena. The damping function, similar to the one used by Tordella and Iovieno 2011 [83], is given by:

$$p(x) = \frac{1}{2} \left[1 + tanh\left(a\frac{x}{L_x}\right) tanh\left(a\frac{x - L_x/2}{L_x}\right) tanh\left(a\frac{x - L_x}{L_x}\right) \right] \tag{2.4}$$

and permits to tune the initial thickness of the interfacial region and hence the steepness of the gradients thereby. Different choices of a have been tried and in the end a value of $a = 20\pi$ has been used. In particular, it has been noticed that the choice of high values of parameter a introduces sharp and persistent peaks of vorticity at the interface during all the propagation. As most basic example no mechanism for sustaining turbulence is introduced in the simulation of the propagation, which means that after the initial homogeneous isotropic condition is produced and the interface introduced, no further forcing is provided during the propagation run. In doing this, any possible influence from the forcing mechanism is prevented, producing a flow where turbulence propagates and decays at the same time. Following such a procedure, over 20 independent realizations have been produced and statics have

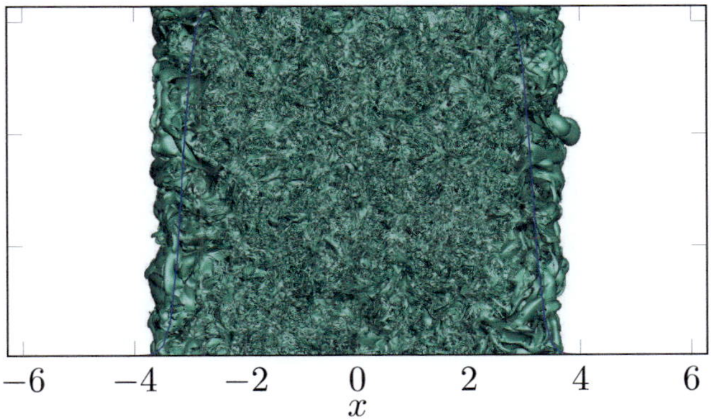

Figure 2.2: The smoothing function from Equation 2.4 superimposed over plotted iso-surfaces of the enstrophy field.

Table 2.1: Initial parameters of the simulation: l_0 is the integral length scale of the initial condition and Δ_x/η_0 the initial resolution.

$L_x \times L_y \times L_z$	Δt	$1/Re$	l_0	λ_0	Re_{λ_0}	η_0	$\Delta x/\eta_0$
$4\pi \times 2\pi \times 2\pi$	$5 \cdot 10^{-5}$	0.005	0.4	0.19	120	0.0075	1.6

been ensemble averaged among these realizations. The position of the average interface position is tracked in time by locating the outermost points where a minimum level of enstrophy is reached for each y, z coordinate. This threshold level has been set to $0.02\Omega_b$ or 2% of the average enstrophy in the midplane of the turbulent flow (from now referred as 'bulk') at a given time. Thus, the threshold decreases in time with the same rate as the turbulence decay of the bulk, it will be shown as this definition of thresholds permits to effectively detect the region of enstrophy growth in this time and spatially evolving flow. After an initial transient, the interface position defined in such a way shows a growth rate which is proportional to $\sqrt{t}$, which is typical to for time evolving shearless turbulent fronts [42]. Statistics are sampled only after this growth rate is established at around 5 integral times scales from the beginning of the decay, while $Re_\lambda = 50$.

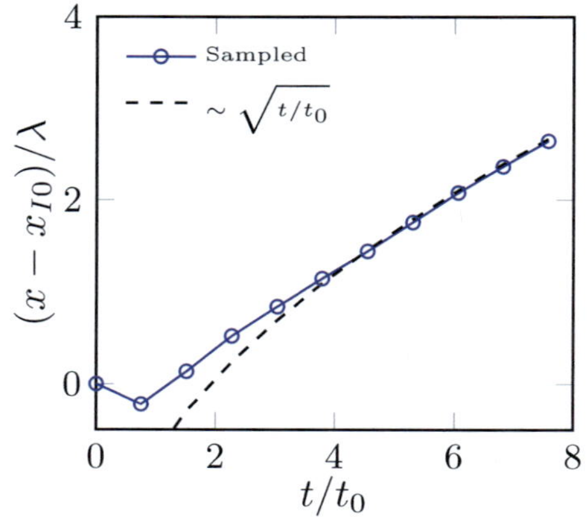

Figure 2.3: Average interface position with respect to time. x_{I0} denotes the position of the interface at the beginning of the decay. Adapted from [14].

2.2.1 Enstrophy budgets for a planar symmetric turbulent front

The purpose of the following analysis is to gain a better understanding of which contributes are more relevant to the propagation of the turbulent front and at which scales each contribute is most active. The decaying flow is constituted by a bulk of turbulent flow, which remains quasi-homogeneous and isotropic for all the duration of the simulation. Further away in $x-$direction from the center of the bulk the flow grows more and more inhomogeneous. Statistical homogeneity in $y - z$ direction is nevertheless maintained and such planar homogeneity permits us to reduce equation 2.1 to:

$$\frac{\partial \langle \Omega \rangle}{\partial t} = \langle \omega_i \omega_j s_{ij} \rangle - \nu \left\langle \frac{\partial \omega_i}{\partial x_j} \frac{\partial \omega_i}{\partial x_j} \right\rangle - \frac{\partial \langle \Omega u \rangle}{\partial x} + \nu \frac{\partial^2 \langle \Omega \rangle}{\partial x^2}, \qquad (2.5)$$

where$\langle \Omega \rangle = \langle \omega_i \omega_i \rangle / 2$ and $\langle \cdot \rangle$ indicate both the ensemble average and spatial average in the homogeneous $y - z$ planes. Here it can be seen that the major differences from a completely homogeneous isotropic turbulent flow reside in the fact that, in general, the diffusive and advective terms (respectively $-\frac{\partial \langle \Omega u \rangle}{\partial x}$

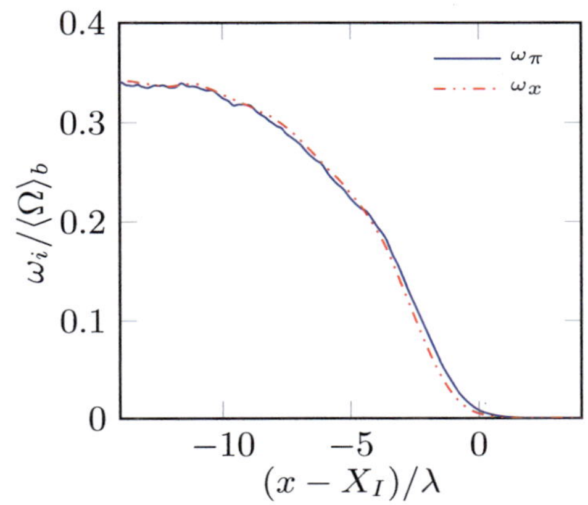

Figure 2.4: Average magnitude of the vorticity components ω_p in the homogeneous plane and ω_x normal to it as a function of the distance from the average interface position. The value are normalized by $\langle\Omega\rangle_b$ the average magnitude of the total vorticity in the bulk. Adapted from [14].

and $\nu\frac{\partial^2\langle\Omega\rangle}{\partial x^2}$) are in general non-zero on average. Both terms contribute only through their derivatives in the inhomogeneous direction already pointing out as this inhomogeneity is required in order to arise the spatial fluxes necessary to the propagation of the front. The presence of the interface hence leads to both, the usual redistribution of fluctuations throughout the space of scales as well as into the physical space. Again it can be observed as both the inertial and viscous terms redistribute enstrophy both into the physical space ($-\frac{\partial\langle\Omega u\rangle}{\partial x}$ and $\nu\frac{\partial^2\langle\Omega\rangle}{\partial x^2}$) and into the scale space ($\langle\omega_i\omega_j s_{ij}\rangle$ and $\nu\left\langle\frac{\partial\omega_i}{\partial x_j}\frac{\partial\omega_i}{\partial x_j}\right\rangle$) hence once again this to process remain entangled in the two spaces.

Figure 2.5 shows the various terms of the enstrophy equations in function of the distance x from the average position of the interface X_I and normalized by the Taylor microscale λ measured in the bulk of the flow. In the analysis the flow is divided in three regions, identifiable from Figure 2.5: firstly the *bulk region* for $(x - X_I)/\lambda) < 10$ is shown. There turbulence is in good approximation still homogeneous and isotropic, the total variation of enstrophy is dominated by the production and the viscous dissipation sensibly in favour of the latter. These two terms are mostly producing, redistributing and destroying enstrophy in the

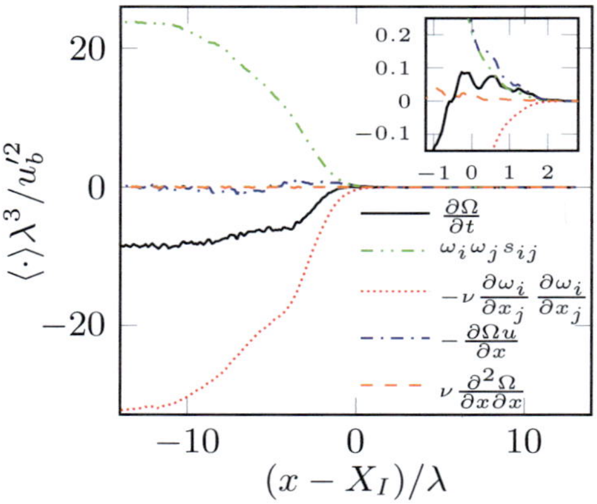

Figure 2.5: Components of the enstrophy equations as a function of the distance from the average interface position. The inset is a magnification of the same plot at the interface region. Adapted from [14].

scales space. The negligible contribution of the spatial fluxes (both diffusive and advective) denotes that, in the physical space, enstrophy is redistributed roughly uniformly in all the directions within this region, the presence of the interface apparently has little effect on enstrophy dynamics in the *bulk region*. Also, the importance of the balance between dissipation and production mechanics in a turbulent flow is shown. As a matter of fact, all the other terms of the enstrophy budget are some orders of magnitude smaller and the time variation of enstrophy in most of the flow appears to be determined only by the difference between $\langle \omega_i \omega_j s_{ij} \rangle$ and $\langle \nu \partial \omega_i / \partial x_j \partial \omega_i / \partial x_j \rangle$.

Moving forward towards the interface, it is possible to identify a region around $-10 < (x - X_I)/\lambda < -0.5$ that from here on will be defined *inhomogeneous layer* due to the growing effect of the inhomogeneous gradient of enstrophy in the flow. Following the gradual reduction of enstrophy inside this layer, both production and dissipation decrease in magnitude. Most notably, the increasing mean gradient of enstrophy is accompanied by a non-negligible contribution from the advective flux. Initially, the negative flux draws enstrophy from the region $-10 < (x - X_I)/\lambda < -5$, then it release enstrophy in the region $-5 < (x - X_I)/\lambda < -0.5$ where the advective flux becomes positive. Enstrophy advection reaches its peak at $(x - X_I)/\lambda = -2$ and for most of the *inhomogeneous layer* it remains orders of magnitude larger than the viscous diffusion.

Finally, the *interfacial layer* is reached for $(x - X_I)/\lambda > -0.5$. In this layer the intensity of the advective flux becomes comparable to the one of $\omega_i \omega_j s_i j$, the joint positive contribute of these two terms is strong enough to overcome the viscous dissipation and to give rise to a positive total variation of enstrophy in time $\partial \langle \Omega \rangle / \partial t$. Also notable is a weak but positive viscous diffusion which marks the presence of the viscous superlayer in this region. Despite being small, the growth rate still allows the propagation of the turbulent front in face of the general decay of turbulence in the rest of the flow. It must be stressed that this growth is driven by the inviscid transport of enstrophy via velocity fluctuations and cannot be addressable to diffusion alone. Also notable is the fact that the regions defined here are independent from the time during decay, hence the Re_λ, at which they are considered considered. Comparing Figures 2.5 and 2.4 it is possible to see how the regions where the advective flux is more intense also coincides with those regions where the anisotropy of vorticity is stronger. The negative advection close to the bulk is thus affecting mainly the in-plane component of vorticity, while its increase close to the interface can be attributed to tilting of out-of-plane vorticity forced by the vicinity of the interface itself.

2.2.2 Spectral Enstrophy Budget

If a more direct analysis of the scales dynamics is sought for, one way to investigate it is to analyse the spectral content of the components of the enstrophy. Due to the inhomogeneity in $x-$direction, the enstrophy spectrum will depend both on the location in the physical space and on the wave-numbers. In particular, for the symmetries of the problem, it is possible to reformulate the enstrophy written in the wavenumber space along the homogeneous (y, z)-directions and in physical space along the inhomogeneous x-direction, i.e. $\tilde{\Omega} = \tilde{\Omega}(x, k_\pi)$ where $\tilde{(\cdot)}$ refers to a 2D Fourier transform in the homogeneous $(y - z)$-space and $k_\pi = k_{y,z}$.

This 2D spectrum of enstrophy is depicted in Figure 2.6 normalised by $\lambda^3 / u_b'^2$ and as a function of the distance from the interface and the wave-number $k\lambda$. For two different times during the decay they show how the maximum enstrophy

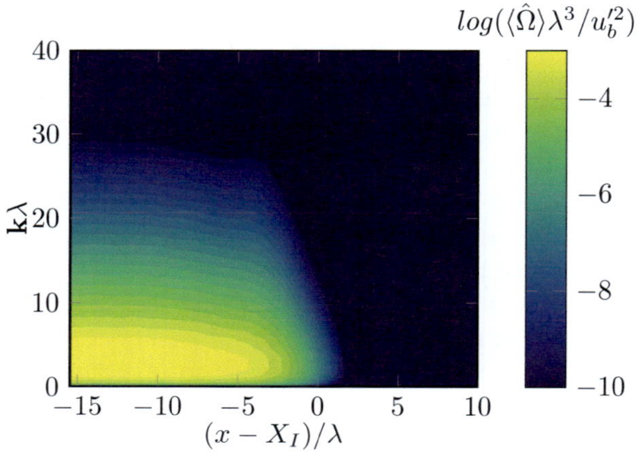

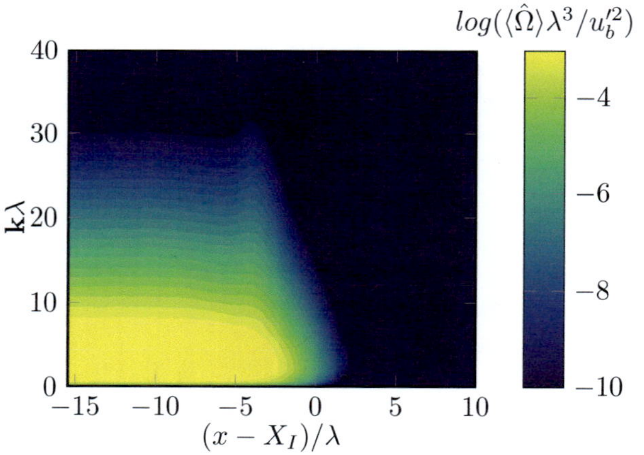

Figure 2.6: Isocontours of $log(\langle\hat{\Omega}\rangle\lambda^3/u_b'^2)$ in the $(k, x - X_I)$-space for (a) $t/t_0 = 4.5$ and (b) $t/t_0 = 8$. Adapted from [14].

is located in the bulk at a $k\lambda$ of around 2.5, this correspond to a scale in the physical of $L/\lambda = 2\pi/(k\lambda) \approx 2.51$. Noteworthy is how the spectral distribution of enstrophy remains rougly constant within all the *bulk region*, only into the *inhomogeneous layer* a gradual reduction of the enstrophy reduction of enstrophy appears and becomes more intense getting closer to the interface.

This region of enstrophy depletion corresponds to the region of Figure 2.5 at $-10 < (x-X_I)/\lambda < -5$ where the advection acts as a sink. Similarly the positive advection observed in Figure 2.5 here gives rise to an increase of enstrophy in the outer part of the inhomogeneous layer for $-5 < (x - X_I)/\lambda < -0.5$ at intermediate to high wave-numbers $k\lambda$. This is particularly evident at $t/t_0 = 8$, where the peak reaches even higher wave-numbers than in the *bulk region*, while small wave-numbers see a decay of enstrophy in both cases. The *interfacial layer* sees a general decay of enstrophy at all wave-numbers, with a depletion of the in-plane small scales which appears almost linear in space. For what concerns the evolution of the spectral enstrophy in this hybrid Fourier-physical space one should start from its formulation in the Fourier space only. The enstrophy in the Fourier space is $\hat{\Omega} = \hat{\omega}_i\hat{\omega}_i^*/2$, where $\hat{\cdot}$ denotes the Fourier transform and * denotes the complex conjugate. In the case of homogeneous isotropic turbulence, the balance equation for spectral enstrophy is:

$$\frac{\partial \hat{\Omega}}{\partial t} = -ik_j\hat{\omega}_i^*\widehat{\omega_i u_j} + \hat{\omega}_i^*\left(\widehat{\omega_j\frac{\partial u_i}{\partial x_j}}\right) - 2\nu k^2\hat{\Omega}, \tag{2.6}$$

where $k^2 = k_x^2 + k_y^2 + k_z^2$ and i is the imaginary unit. As for the spectral enstrophy alone the budget equation due to the inhomogeneity in $x-$direction will actually depend both, on the location in the physical space and on the wave-numbers. Considering this inhomogeneity, the resulting equation will be:

$$\frac{\partial \tilde{\Omega}}{\partial t} = \underbrace{-ik_\pi\tilde{\omega}_i^*\widetilde{\omega_i u_\pi}}_{T_k} \underbrace{- \tilde{\omega}_i^*\frac{\partial \widetilde{\omega_i u}}{\partial x}}_{T_x} + \underbrace{\tilde{\omega}_i^*\left(\widetilde{\omega_j\frac{\partial u_i}{\partial x_j}}\right)}_{\gamma} \underbrace{- 2\nu k_\Pi^2\tilde{\Omega}}_{\varepsilon_k}$$
$$+ \underbrace{\nu\frac{\partial^2\tilde{\Omega}}{\partial x^2}}_{D_x} \underbrace{- \nu\frac{\partial\tilde{\omega}_i}{\partial x}\frac{\partial\tilde{\omega}_i^*}{\partial x}}_{\varepsilon_x}, \tag{2.7}$$

where $k_\Pi^2 = k_y^2 + k_z^2$. Equation (2.7) allows us to analyse the dynamics of enstrophy in both, the wavenumber and the physical space. Since the wavenumber space k_π is isotropic, the integral of Equation (2.7) over a shell in the (k_y, k_z)-space of radius k and thickness dk is considered. In such a way Equation (2.7) turns to be a function only of k and of the position x. Compared to Equation 2.1, the new coordinate k adds a further dimension to the budget by showing the contributes from advection, diffusion, production and dissipation at different wave-numbers. Thus, while the scale transfer in Equation 2.1 could only be inferred here the wave-number dependence is made explicit. This formulation is only representative of the scales in $y - z$ directions, while the information about the scales in $x-$direction at best can only be deduced. Looking into the details of Equation 2.7, it is possible to divide it in contributes into the physical space and contributes into the Fourier space. γ is the production term due to vortex stretching. Here, as in the conventional budget, it represents a source of enstrophy due to non-local interactions of vorticity and strain, it should not surprise therefore that in Equation 2.7 this only has contributes in the Fourier space. The dissipative term in its spectral form gives rise to two different contributes, ϵ_k and ϵ_x, which are related respectively to the gradients into the in-plane wave-numbers k and to the spatial gradients in $x-$direction. Similarly there are a spectral flux T_k which redistribute enstrophy among different in-plane wave numbers and an inertial spatial flux that transfers enstrophy towards different spatial locations in $x-$direction. The last flux is the viscous diffusion in the physical space.

The analysis of the spectral budget can at some points be simplified, if the terms from Equation 2.7 are regrouped in a spatial flux $S_x = T_x + D_x$, also the dissipative terms and the production can be grouped into an effective source term $\xi = \gamma + \epsilon_k + \epsilon_x$. In this way Equation 2.7 can be rewritten as:

$$\frac{\partial \tilde{\Omega}}{\partial t} = \xi + S_x + T_k, \tag{2.8}$$

which immediately describes, as a function of k and the distance from the interface $(x - X_I)/\lambda$, how enstrophy is generated/destroyed and transferred into both, the physical space and both the wave-number space. Here the analysis proceeds following the subdivision introduced in the previous section studying the spectral budgets of the *bulk region* in the following order: the inner and the outer part of the *inhomogeneous layer* finishing with the *interfacial layer*.

As it has previously been done with the conventional budget, the *bulk region* is first considered as this region behaves in good approximation like a decaying homogenous isotropic turbulence. Indeed, looking at the time variation of enstrophy in Figure 2.7, it can be seen how enstrophy is decaying over the whole spectrum. The spectral flux T_k, here negative at small $k\lambda$ and positive

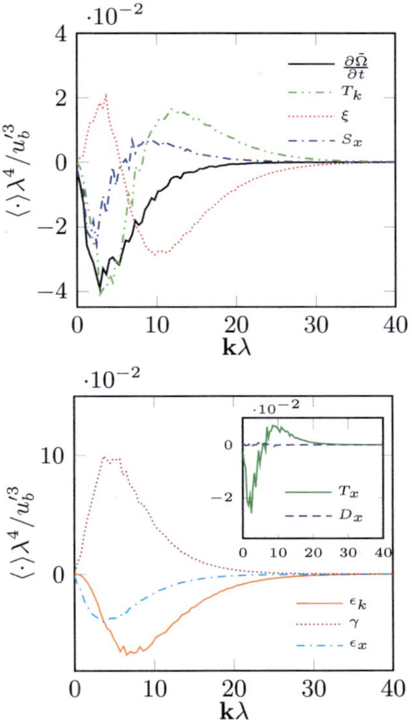

Figure 2.7: Spectral enstrophy budget within the bulk turbulent region for $t/t_0 = 4.3$. The terms of Equation (2.8) are shown in (a) while the different components of ξ and S_x from Equation (2.7) are delineated in (b). Adapted from [14].

at large ones, represents the classical view of a direct cascade of enstrophy from the large scales ($k\lambda < 7$) towards the smaller ones ($k\lambda > 7$). The prevalent negativeness of the source term ξ means that the dissipative terms are stronger at almost all the wave-numbers and only at $k\lambda < 7$ the production term can overcome the dissipation. Despite the flow is homogeneous in the *bulk region*, one can observe as a non-negligible S_x and ϵ_x arise here. These show a maximum magnitude of about one half of their spectral counterparts T_k and ϵ_k and in this region they can be interpreted as respectively the spectral transfer and dissipation in the x−scales space (or wave-numbers k_x). This is confirmed by the fact that the integral $\int S_x dk$ here is negligible, hence it does not produce any net transfer of enstrophy towards the interface but it is only cascading it from low k_x to larger ones. In a similar way, in the homogeneous turbulent *bulk*, ϵ_k is the result of dissipation of enstrophy in the k_x space. This is again confirmed by the fact that $\int \epsilon_x dk \approx \int \epsilon_k dk/2$ which is what is expected from isotropic turbulence. It will be shown how, getting closer to the interface, the behaviour of the spatial terms S_x and ϵ_x will deviate from what was observed here under the effect of the growing inhomogeneity in the flow. There these terms will be less and less representative of the underlying scale transfer and will become predominantly determined by the spatial mean gradients. In fact, it is possible to observe a departure from the homogeneous behaviour of the *bulk region* already in the inner part of the *inhomogeneous layer*. The spectra of a representative section of this region are depicted in Figure 2.8 for $(x - X_I)/\lambda = -7$, a section which corresponds to the location where the enstrophy drain due to inertial advection takes place in Figure 2.5. Here, as in the *bulk*, enstrophy is decaying at all the scales. On the qualitative level, here the spectra show the same general behaviour as in the *bulk* with the only noticeable distinction of the spatial flux S_x, which is a sink for a slightly large range of wave-numbers compared to how it was in the *bulk*.

Figure 2.9 shows how the spectra radically change in the outer part of the *inhomogeneous layer* for $(x - X_I)/\lambda = -2$. In this region enstrophy is being released by the inertial advection from the inner textitinhomogeneous layer. It can be observed how, while still negative, the variation of enstrophy is much less intense than in the *bulk* and how the larger wave-numbers are already at equilibrium. The reason for this reduced decay rate is attributable to the now strong spatial flux S_x which is positive at all the scales, implying that enstrophy is being advected at all scales from the inner *inhomogeneous layer* and this flux strongly overcomes the underlying enstrophy cascade in the k_x space. Watching the inset of Figure 2.9, as expected it can be observed how the flux S_x is an essentially inviscid process as the viscous diffusion D_x is still negligible compared to the advection T_x. The diminishing intensity of the enstrophy production γ preventsto overcome the dissipative terms, ϵ_x and ϵ_k at all the scales, and that is in such a way that the source ξ is negative everywhere. Also, Figure 2.9 shows that the dissipation ϵ_x due to the gradients in the inhomogeneous direction is growing in magnitude when compared to ϵ_k and it peaks at the same numbers as the production term γ. As in the rest of the flow, the spectral transfer T_k here drains large-scale enstrophy and releases it at smaller ones. The spatial flux S_x is observed to release enstrophy at larger

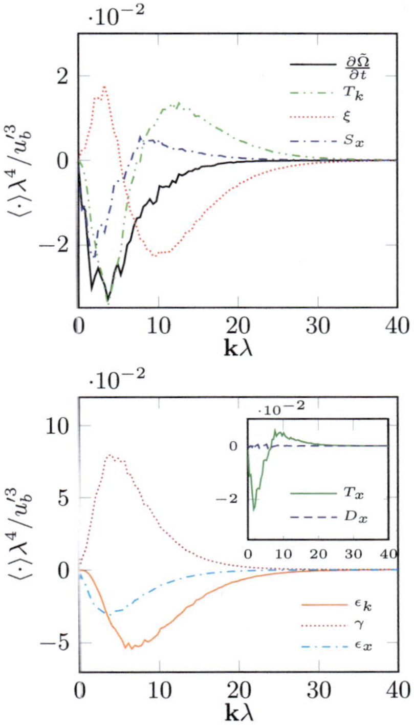

Figure 2.8: Spectral enstrophy budget within the inhomogeneous layer at $(x - X_I)/\lambda = -7$ where the peak of enstrophy draining due to the spatial flux takes place. The terms of Equation (2.8) are shown in (a) while the different components of ξ and S_x from Equation (2.7) are delineated in (b). Adapted from [14].

wave-numbers compared to the one drained in the inner *inhomogeneous layer*. Indeed, while the negative peak of S_x is located at $k\lambda \approx 2$, the positive the flux peaks closer to the interface at $k\lambda \approx 5$. This is in agreement with the representation of S_x as a flux in both the physical and the scales space, as the enstrophy drained at large scales close to the bulk is advected towards the interface and at the same time undergoes through a cascading process feeding enstrophy close to the interface at smaller scales.

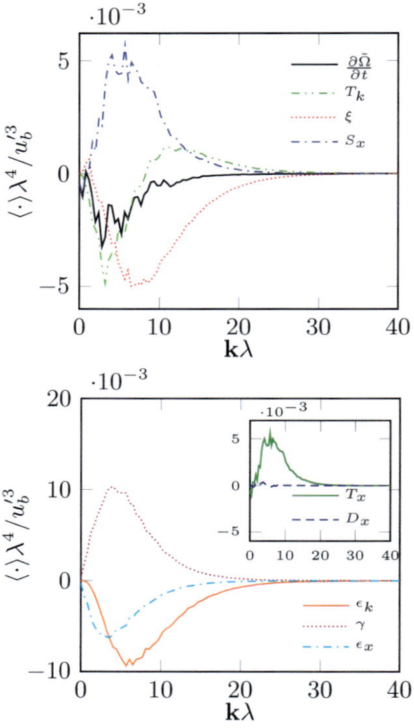

Figure 2.9: Spectral enstrophy budget within the inhomogeneous layer at $(x - X_I)/\lambda = -2$ where the peak of enstrophy source due to the spatial flux takes place. The terms of Equation (2.8) are shown in (a) while the different components of ξ and S_x from Equation (2.7) are delineated in (b). Adapted from [14].

At the interface, depicted in Figure 2.10, it is finally possible to observe a non-negative enstrophy variation at all the wave-numbers. Especially the largest in-plane scales of the flow gain enstrophy while the intermediate to small ones have reached an equilibrium. Again in this generally decaying flow this growth cannot be attributed to the local production of enstrophy γ. Indeed, it remains as in the outer *inhomogeneous layer* always weaker than the dissipation leading to ξ being negative at all wave-numbers. Part of this is due to the increasing relevance of the viscosity at the interface and in particular to the growth of ϵ_x. The growth of enstrophy is again determined by the intensity of the spatial

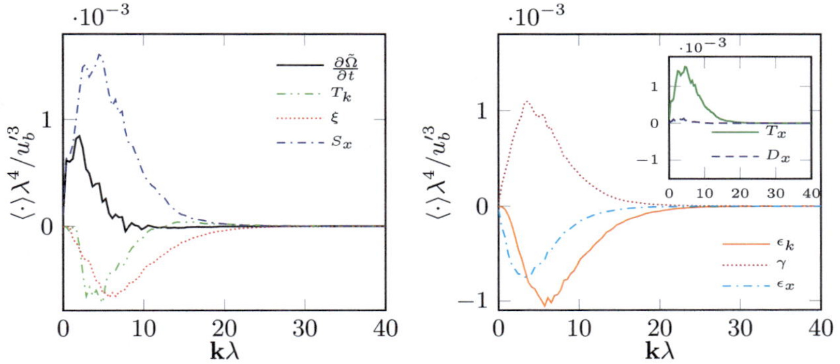

Figure 2.10: Spectral enstrophy budget within the interfacial layer at $(x - X_I)/\lambda = 0$ for $t/t_0 = 4.3$. The terms of Equation (2.8) are shown in (a) while the different components of ξ and S_x from Equation (2.7) are delineated in (b). Adapted from [14].

transport S_x, mostly sustained by the advection T_x. It is also interesting to notice the apparently anomalous behaviour of the spectral transfer T_k that, due to the strong anisotropy, at the interface does not enforce a transfer from small to large wave-numbers anymore, but acts as a sink for all the in-plane scales instead. In accordance with what was pointed out in the previous sections, this is indicative of the spectral transfer of enstrophy due to vortex tilting from the k in-plane space towards the k_x at smaller scales. Finally, the viscous diffusion here as in the budget of Figure 2.5 shows a weak but positive contribute. It is interesting to point out how the in-plane scales of viscous diffusion D_x are even larger than the ones mostly gaining from the spatial transport. This is in apparent contrast with the typical figure of inertial processes being large-scale phenomena and viscous processes being small-scale ones. It must be remembered that D_x is directly representative of diffusion among scales in the in-plane direction, while it has been observed as the out-of-plane scales of viscous diffusion show thicknesses in the order of few Kolmogorov scales. It appears then that diffusion is mostly active in sheets of relatively large in-plane extension and Kolmogorov scale thickness.

As a confirmation of the observations made in the spectral analysis, one can qualitatively assess the scales of the viscous diffusion at the interface. By plotting the viscous diffusion over the surface of the TNTI, as can be seen

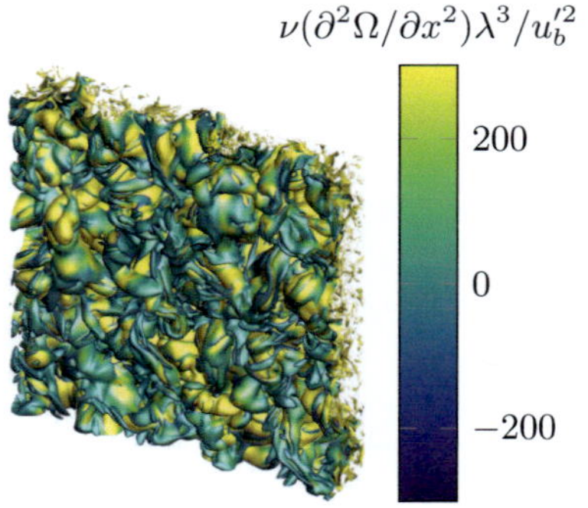

Figure 2.11: Viscous diffusion of enstrophy in a 2D cross-section of the turbulent/non-turbulent interface. Adapted from [14].

in figure 2.11, a qualitative observation reveals that the diffusion is mostly positive in relatively large spots at the interface. The thickness of the viscous diffusion dominated layer observed in previous studies [79], together with the observations from spectral analysis, put together the picture of a thin layer where vorticity fluctuations parallel to the interface diffuse most intensely at large scales. The small-scale nibbling has more the form of a thin but spread vortex sheet than that (often depicted in literature) of a fractal like hierarchy of increasingly small structures that diffuse their vorticity down to the Kolmogorov length scale. Indeed, by observing a cross-section of the field of viscous diffusion $\partial^2\Omega/\partial x_i \partial x_i$ in figure 2.12 as expected we see how viscous diffusion is organized in thin stratified layers of alternating sign with thickness usually around $4 - 5\eta$. The outermost layer being mostly positive is the one which in conditional averages produces the characteristic viscous dominated region known as the viscous superlayer.

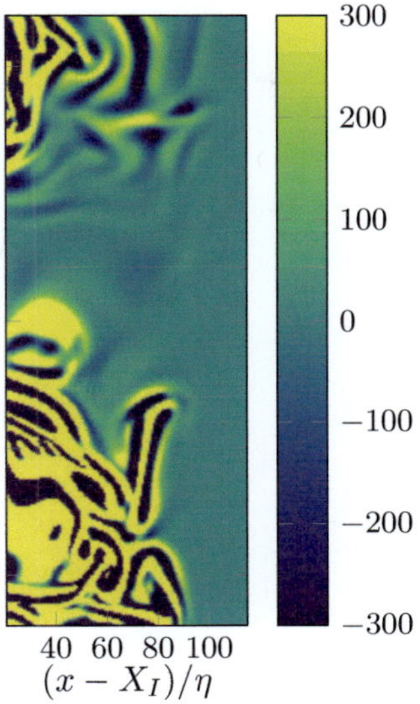

Figure 2.12: Viscous diffusion field in a cross-section of the flow.

3 Dilute polymer solutions models

3.1 Dilute polymer solutions and turbulent flows

In this chapter basics elements on the dynamics of dilute polymer solution will briefly be introduced together with the kinetic theory which is the basis to the most diffuse models for the dynamics of dilute polymer solutions. Polymers are molecules constituted by several repetitions of a same molecular sub-group called monomer. They are ubiquitous in modern industrial applications due to their versatility but are also very common in nature (examples range from the DNA chains to the pectin that thicken jellies and jams). In fluid flow applications polymers gained attention due to some interesting non-Newtonian phenomena that arise in flows of solvents in which they are diluted in small concentrations. The most evident and most studied of such modifications is certainly arise in turbulent bounded flows of dilute polymer solution, where the drag of the flow can be found to be as little as 80% less than equivalent Newtonian flow with matching viscosity [6]. Other anomalous phenomena are for example the dye swelling in jets or rod climbing in rotating tank with a rod in the centre [6, 41].

In dilute polymer solutions the Newtonian linear relation between stress and strain fail to describe the flow for all but the most simple laminar cases. Some viscous-like relations can be used to a certain extent, but generally these fail to describe even qualitative aspects of the flow when it transitions to the turbulent state. In order to better model the rheology of diluted polymer solutions molecular dynamics of the polymer chains have to be accounted for. One of the most successful approaches to the problem is the kinetic theory, which attempts to model polymer molecular dynamics via a stochastic approach. This theory in particular is geared towards modelling the behaviour of high molecular weight, flexible polymers constituted by millions of unbranched chains of a single type of simple monomer (e.g. without phenyl groups, cyclic sugars etc. [41]). Moreover, certain assumptions are required such to limit the applicability of the model to dilute polymer solutions only. "Dilute solution"means a mix of polymers dissolved in a solvent in which the addition of polymers does not significantly alter the viscosity of the solvent: typical concentrations are in the order of few polymer molecules per million solvent molecules [41]. In the polymer chain the single links between the molecules

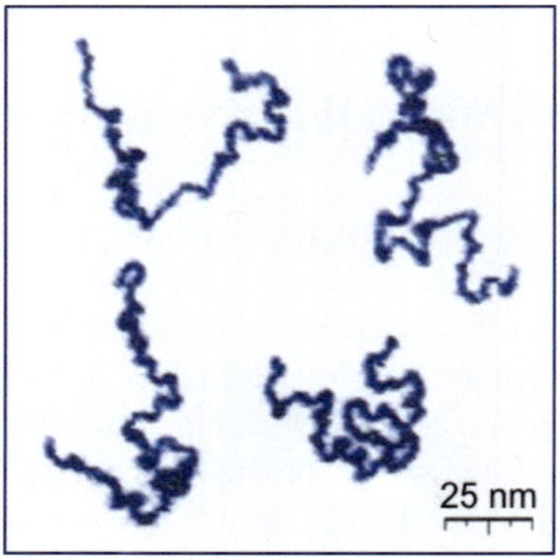

Figure 3.1: An atomic force microscope picture of a poly(2-vinylpyridine) ($n \times C_7H_7N$) chain of about 1500 monomers at different degrees of un-coiling [72].

can rotate and bend allowing a great number of possible configurations of the complete chain as it can be appreciated in Figure 3.1 for poly(2-vinylpyridine) molecules. In absence of shear, the chain tends to assume a compact coiled configuration like the one assumed on the bottom right of Figure 3.1, while under the effect of velocity gradients the different velocities that different parts of the chain experience tend to uncoil the chain stretching it. One of the most typical example in the literature on turbulence of dilute polymer solutions is the *polyethilenoxide* or PEO (monomer $-CH_2 - CH_2 - O$), which can form single chains with up to millions of monomers for an extended length up to the order of few micrometres [41].

For dilute polymer solutions, experimental evidence in laminar flows showed that the rheology of the solution does not show any differences from the one of their solvent [41]. Evident modifications in both the macroscopic and small scale behavior of the flow start to arise only when the flow start to transition to a turbulent state [41]. Evidence shows that, while changing the features of the flow at all scales, polymers more directly interact with the smallest scales of the flow: for example, when drag reduction appears, polymers strongly decrease Reynold stresses in the flow [93], increase the correlation length of small scales and a shift of the energy content towards larger scales compared to a Newtonian flow [87]. Nevertheless, suppression of turbulent properties is not granted in all conditions and in certain instances polymers have been found to actually increase the intensity of turbulent fluctuations [27,54,85]. A direct interaction between small scales of the flow and single polymer chains might look obvious, but it must be remembered that drag reduction in pipe flow experiments manifests itself at Reynolds numbers in the order of 10^3 [41], at these Reynolds numbers in the same experiments the length of all but the heaviest fully stretched polymer molecules are one or two orders of magnitude smaller than the Kolmogorov length scale. Nevertheless, the coupling between the flow and the polymers strongly depends on the coupling between their characteristic scales as the dependence between drag reduction and polymer chain length demonstrate [41]. Hence, in order to compare flows of dilute polymer solutions, the Reynolds equivalence is not enough and an additional non-dimensional group comparing polymer and fluid scales is required. This group usually takes the form of either the Deborah number or the Weissemberg number. The Deborah number $De = \tau/T_o$ where T_o is the characteristic time of the observed phenomena and τ is the polymer relaxation time, i.e., the time required for the shear stress in a simple shear flow to return to zero in constant strain conditions [67]. The Weissemberg number can be defined for shearing flows as the product between the shear rate and the relaxation time such that $Wi = \frac{dU}{dy}\tau$. In a Newtonian fluid under such conditions, the shear stress would immediately go to zero leading to a zero relaxation time while for a Hookean elastic material the stress is maintained indefinitely and the relaxation time is infinite. The Deborah number can be seen as an indicator of how much a material undergoing a deformation with a characteristic time scale T_o behaves as an elastic solid or a Newtonian fluid (with $De = \infty$ for solids and $De = 0$ for fluids). In reality, no perfectly Hookean or Newtonian material exists and all materials show a behaviour in between the two: for example, water has a relaxation time in the order of $10^{-12} s$ and glass one of 28 hours [67]. The relaxation time of a polymer depends on the shape its coils assume in a given configuration and on its orientation with respect to the perturbation. This means that polymer do not have a single relaxation time but a whole spectrum of them. Nevertheless, in many models a single representative relaxation time is usually assumed for simplicity. In the case of dilute PEO-water solution with molecular weight $M_w = 2 \cdot 10^6 g/mol$, the typical relaxation time range is in the order of $10^{-4} s$ depending on the polymer concentration [75].

At the polymer length scale, the smallest velocity gradients in the flow are still perceived as a uniform shear [6], and the reaction of a polymer to these gradients would happen at scales smaller than the Kolmogorov one. Nevertheless, polymers induce macroscopic changes at all scales of the flow. One reason, at least for the case of the PEO, comes to the experimental evidence of the formation of supermolecular structures between several polymer chains even at low concentrations that would form reticulate structures of much larger dimensions [48, 49], but there are other mechanisms with which small-scale inputs can result in the production of large-scale alterations via triadic interactions [87]. As it can be seen from this brief introduction to the topic, polymers increase the complexity of the problem on several levels. The large spectrum of configurations a single polymer chain alone can assume can add millions of degrees of freedom to the problem and a single fluid parcel would contain several of them. One is then forced to resort to a number of simplifications in order to reach a numerically treatable formulation of the problem. In this process a number of dependencies on the properties of the molecules employed, on their interaction with the surrounding solvent and other polymer's molecules, all characteristics of the solution will be deliberately over-simplified or ignored. The resulting models have the advantage of qualitatively capturing some of the relevant dynamics of turbulent flows of dilute polymers solution. Moreover, as for turbulence modelling, the success of a simple model can help to direct the research in understanding what is mostly relevant in the observed physical phenomenon. Thus, in the following the theory behind the FENE-P model used in the rest of this work will be introduced.

3.2 The kinetic theory

The most diffuse models in computational rheology of dilute polymer solutions start from the kinetic theory in order to reach a continuum formulation of the relation, or constitutive equation, between stress and deformation at a macroscopic level, .i.e. at the level of the fluid element. The kinetic theory is a coarse-grained model for the polymer conformation, which means it does not try "to provide a description of the fluid at a molecular leveländ that "processes at atomistic levels are ignored" [67]. The contribution of the conformation of the polymers molecules to the stress in a fluid particle is represented via the ensemble average of the conformation of a number of modellised polymers. The typical basic unit of such models is composed by two spherical bodies, or

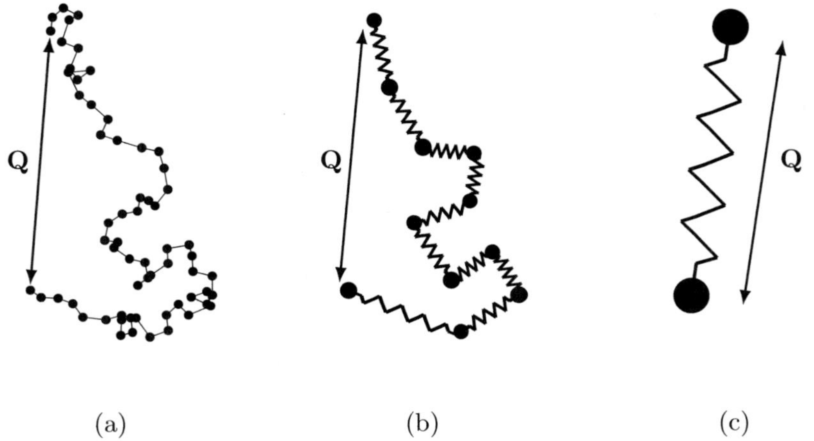

Figure 3.2: Various degrees of simplification of the polymer model with its end-to-end vector Q. (a) Multi chain dumbbell model with rigid connectors between the beads. (b) Multi-chain dumbbell model with elastic connectors. (c) Single elastic dumbbell model.

beads, that are linked together. Due to its shape it is also referred to as 'dumbbell model'. At an atomistic level the monomers are linked together with fixed bond angles, bond lengths and torsion angles. It is possible to reduce the complexity of the model by substituting concatenations of monomers over which a certain correlation in the orientation persists with a single dumbbell element. Such correlation length is called persistence length and is a measure of the degree of flexibility of a molecule. On distances shorter than this persistence length, the molecule behaves as a flexible elastic rod, while at larger distances the correlation vanishes and the molecule will behave as a freely jointed chain made of rigid or elastic links with completely flexible joints [70]. In the model, the continuously distributed mass of the polymer molecule is assumed to be distributed over a finite number of discrete beads at the junction points of the chain. The beads not only account for the mass but are also the nodes on which the external forces can act, while in the kinetic theory the links represent the internal reaction forces due to the chemical bonds. Depending on how coarse the model is, the basic element can represent a limited subset of monomers of the chain up to the whole polymer molecule, as depicted in Figure 3.2. Multi-chain models, thought simplified, still require the addition of several hundred degrees of freedom per point of the velocity field. With the computational power currently available, those are applicable to only very simple flow cases. With the purpose of studying relatively complex turbulent flows, here a single dumbbell model will be used. It will be shown how this approach limits the additional degrees of freedom to six, introducing six more equations to the Navier-Stokes system and maintaining a continuum formulation of the problem. In the following part the derivation of the Finite Extensibility Nonlinear Elastic (FENE) model is demonstrated following mostly Owens and Phillips [67]. An

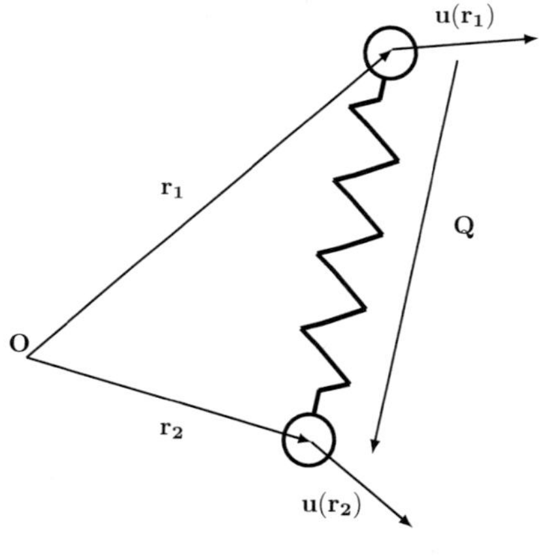

Figure 3.3: Single dumbbell model constituted of two beads linked by an elastic force.

elastic dumbbell immersed in a Newtonian solvent is considered as shown in Figure 3.3. Such dumbbell consists of two beads with mass m and position vectors $\vec{r}_1$ and $\vec{r}_2$ relative to some fixed coordinate system. The equations of motion for the beads in the dumbbell are:

$$m\frac{d}{dt}\Big(\frac{d\vec{r}_i}{dt} - \vec{u}(\vec{r}_i)\Big) = -\zeta\Big(\frac{d\vec{r}_i}{dt} - \vec{u}(\vec{r}_i)\Big) + \vec{F}_i + \vec{B}_i \tag{3.1}$$

for $i = 1, 2$. Here $\vec{u}(\vec{r}_i)$ is the velocity of the solvent at position $\vec{r}_i$ of the ith bead, $\vec{F}_i$ is the force on the ith bead exerted by the spring and $\vec{B}_i$ is the Brownian force due to the impact of the solvent molecules on the ith bead. The constant ζ is the friction coefficient and it arises from the Stoke's law: the drag force on the ith bead is assumed to be directly proportional to the difference between the bead velocity and that of the surrounding medium. With the hypothesis of spherical beads, the friction coefficient is $\zeta = 6\pi\eta_s a$, where a is the radius of the bead, and η_s is the solvent viscosity. In modelling solvent-beads interaction in such a way, hydrodynamic interactions, i.e, the effect that one bead may have on the velocity of the solvent in the vicinity of the other bead, are neglected. External inertial forces and weak chemical bound forces, as the ones between solvent and beads or beads with other beads, are neglected as well. The characteristic time scale of the bead velocity fluctuations due to the Brownian forces is represented by the ratio $\lambda_B = m/\zeta$. When the considered time scales of the flow are large compared to λ_b, the Brownian force $\vec{B_i}$ can be written in the form

$$\vec{B_i} = \sqrt{2kT\zeta}\frac{d\vec{W_i}}{dt} \tag{3.2}$$

where $\vec{W_i} = \vec{W_i}(t)$ is a multi-dimensional Wiener process, k is the Boltzmann constant and T the temperature.

Given that Wiener processes are Gaussian-stochastic processes, are completely characterized by the mean and auto-correlation of its components $W_{i,j}$:

$$E(W_{i,j}(t)) = 0, \tag{3.3}$$

$$E(W_{i,j}(t)(W_{i,j}(t')) = min(t,t'). \tag{3.4}$$

The coefficient $\sqrt{2kT\zeta}$ is derived by the principle of equipartition of energy. This states that, for a system in equilibrium, the kinetic energy associated with each physical component of the velocity $d\vec{r_i}/dt - \vec{u}(\vec{r_i})$ for the ith bead is $kT/2$.

The velocity of the flow can be expressed as the truncated Taylor expansion

$$\vec{u}(\vec{r_i}) = u(\vec{0}) + \nabla\vec{u}^T\vec{r_i}. \tag{3.5}$$

If $\vec{Q} = \vec{r_2} - \vec{r_1}$ denotes the end-to-end vector of the dumbbell, the two velocity components can be subtracted in order to introduce the relative velocity

$$\vec{V} = \frac{d\vec{Q}}{dt} - \nabla\vec{u}^T\vec{Q}. \tag{3.6}$$

In order to reformulate Equation 3.1 as a function of the relative velocity and obtain the first-order system of equations

$$m d\vec{V} = -(\zeta\vec{V} + 2\vec{F})dt + 2\sqrt{kT\zeta}d\vec{W}_t, \tag{3.7}$$

$$d\vec{Q} = (\vec{V} + \nabla\vec{u}^T\vec{Q})dt. \tag{3.8}$$

Here $\vec{F} = \vec{F}_1 = -\vec{F}_2$ and $\vec{W}_t = (\vec{W}_2 - \vec{W}_1)/\sqrt{2}$. These equations are a system of stocastic differential equation of a Ito processes, the first of which is driven by the Wiener process $\vec{W}_t$. The probability that a dumbbell has an orientation $\vec{Q}$ to $\vec{Q} + d\vec{Q}$ and a velocity in the range$\vec{V}$ to $\vec{V} + d\vec{V}$ at time t is given by $\mathbf{\Psi}(\vec{Q}, \vec{V}, t)d\vec{Q}d\vec{V}$ where $\mathbf{\Psi}(\vec{Q}, \vec{V}, t)$ is the dumbbell probability density function. Then the equation that describes the time evolution of the probability density function is the Fokker-Planck equation governing $\mathbf{\Psi}(\vec{Q}, \vec{V}, t)$ which is

$$\frac{\partial\mathbf{\Psi}}{\partial t} = -\frac{\partial}{\partial\vec{Q}}\left[(\vec{V} + \nabla\vec{u}^T\vec{Q})\mathbf{\Psi}\right] + \frac{1}{m}\frac{\partial}{\partial\vec{V}}\left[(\zeta\vec{V} + 2\vec{F})\mathbf{\Psi}\right] + \frac{2kT}{m^2}\frac{\partial\mathbf{\Psi}^2}{\partial\vec{V}^2}. \tag{3.9}$$

In this equation the first two right-hand-side terms determine the drift in time of the distribution, while the third right-hand-side ones determine its diffusion. The reaction force $\vec{F}$ will be considered to be an entropic spring force law. In such law, the extension of a polymer chain reduces the configuration space of the polymer and its entropy, the restoring force arises by the tendency of the chain to return to a higher entropy state when the extension force is removed (Treloar (1975), Physics of Rubber Elasticity). Such restoration force is represented by the law

$$\vec{F} = Hf(Q)\vec{Q}, \tag{3.10}$$

where H is a spring constant and $f(Q)$ a scalar function of the dumbbell length $Q = |\vec{Q}|$. The introduction of the relaxation time scale of the dumbbell as $\lambda_1 = \zeta/4H$ leads to

$$\frac{\partial\mathbf{\Psi}}{\partial t} = -\frac{\partial}{\partial\vec{Q}}\left[(\vec{V} + \nabla\vec{u}^T\vec{Q})\mathbf{\Psi}\right] + \frac{1}{\lambda_b}\frac{\partial}{\partial\vec{V}}\left[\left(\vec{V} + \frac{1}{2\lambda_1}f(Q)\vec{Q}\right)\mathbf{\Psi}\right] + \frac{2kT}{m\lambda_b}\frac{\partial\mathbf{\Psi}^2}{\partial\vec{V}^2}. \tag{3.11}$$

A contraction of the Equation 3.11 is seeked in such a way that it is defined for the marginal probability density function of the end-to-end vector only

$$\psi(\vec{Q}, t) = \int_{\vec{V}} \mathbf{\Psi}(\vec{Q}, \vec{V}, t)d\vec{V}. \tag{3.12}$$

In order to do so, first Equation 3.11 is integrated with respect to $\vec{V}$, obtaining the continuity equation

$$\frac{\partial \mathbf{\Psi}}{\partial t} = -\frac{\partial}{\partial \vec{Q}}\left[(\ll \vec{V} \gg +\nabla \vec{u}^T \vec{Q})\mathbf{\Psi}\right], \tag{3.13}$$

where the velocity-space average $\ll \cdot \gg$ is defined as

$$\ll \cdot \gg = \frac{1}{\psi}\int_{\vec{V}} \cdot\, \mathbf{\Psi}(\vec{Q},\vec{V},t)d\vec{V}. \tag{3.14}$$

Then Equation 3.11 is multiplied by $\lambda_b \vec{V}^T$, integrated it with respect to $\vec{V}$ and letting $\lambda_b \to 0$ it leads to

$$\lambda_b \frac{\partial}{\partial \vec{Q}} \cdot \left(\ll \vec{V}\vec{V}^T \gg \psi\right) + \ll \vec{V} \gg + \frac{1}{2\lambda_1} f(Q)\vec{Q}\psi = \mathbf{0}. \tag{3.15}$$

Thirdly, multiplying Equation 3.11 by $\lambda_b \vec{V}\vec{V}^T$ and again integrating it with respect to $\vec{V}$ and letting $\lambda_b \to 0$ it gives us the Maxwell-Boltzmann relation for the kinetic energy of the dumbbell in equilibrium:

$$\frac{1}{2} m \ll \vec{V}\vec{V}^T \gg = kT\mathbf{I}. \tag{3.16}$$

Combining Equations 3.14, 3.15 and 3.16, we finally obtain the contracted Fokker-Planck equation

$$\frac{\partial \psi}{\partial t} = -\frac{\partial}{\partial \vec{Q}}\left[\nabla \vec{u}^T \vec{Q}\psi - \frac{1}{2\lambda_1} f(Q)\vec{Q}\psi - \frac{2kT}{\zeta}\frac{\partial \psi}{\partial \vec{Q}}\right]. \tag{3.17}$$

We may define an ensemble average $\langle \cdot \rangle$ for any function g of $\vec{Q}$ by

$$\langle g(\vec{Q})\rangle = \int_{\mathbb{R}^3} g(\vec{Q})\psi(\vec{Q},t)d\vec{Q}, \tag{3.18}$$

and relate the extra-stress tensor $\mathbf{T}$ to the ensemble average of the dyadic product $\vec{Q}\vec{F}$

$$\mathbf{T} = -nkT\mathbf{I} + \eta_s \dot{\gamma} + nH\langle \vec{Q}\vec{Q}^T f(Q)\rangle, \tag{3.19}$$

where n is the number density of the dumbbells. Equation 3.19 is also called the Kramer form of the stress tensor. In order to obtain a constitutive relation for the extra-stress $\mathbf{T}$, we multiply Equation 3.17 by $\vec{Q}\vec{Q}^T$ and integrate it over $\mathbb{R}^3$. Using the divergence theorem and assuming that $\psi \to 0$ as $|\vec{Q}|$ tends to its maximum permissible length we obtain

$$\frac{\partial}{\partial t}\langle \vec{Q}\vec{Q}^T \rangle - \nabla\vec{u}^T \langle \vec{Q}\vec{Q}^T \rangle - \langle \vec{Q}\vec{Q}^T \rangle \nabla\vec{u} = \frac{4kT}{\zeta}\mathbf{I} - \frac{1}{\lambda_1}\langle \vec{Q}\vec{Q}^T f(Q)\rangle. \qquad (3.20)$$

The left-hand side of Equation 3.20 is called upper-convected derivative of $\langle \vec{Q}\vec{Q}^T \rangle$, and is denoted by $\overset{\triangledown}{\langle \vec{Q}\vec{Q}^T \rangle}$.

3.3 The FENE model and the Peterlin closure

The Finite Extensibility Non-linear Elastic (FENE) model us as connector force the law

$$\vec{F} = \frac{H\vec{Q}}{1 - (Q^2/Q_0^2)}, \qquad (3.21)$$

where $Q^2 = tr(\vec{Q}\vec{Q}^T)$ and Q_0 is some finite constant. Such force law prevents the spring to be extended beyond the length Q_0. The problem with such connector force lays in the treatment of the term $\langle \vec{F}\vec{Q}^T \rangle = \langle \vec{Q}\vec{Q}^T f(Q)\rangle$, which implies knowing the ensemble average of the product $\vec{Q}\vec{Q}^T f(Q)$ afore-hand. The problem has been by-passed by using relation [6] as connector force instead

$$\vec{F} = \frac{H\vec{Q}}{1 - \langle Q^2/Q_0^2 \rangle}, \qquad (3.22)$$

in such a way that when the ensemble average of the diatic product $\langle \vec{F}\vec{Q}^T \rangle$ is plugged into the Kramer expression 3.19 we obtain

$$\mathbf{T} = \tau + \eta_s\dot{\mathbf{S}} = -nkT\mathbf{I} + \frac{nH\langle \vec{Q}\vec{Q}^T \rangle}{1 - \langle Q^2/Q_0^2 \rangle} + \eta_s\dot{\mathbf{S}}. \qquad (3.23)$$

The final relation for the evolution of the conformation tensor using the FENE-P model is

$$\frac{\partial}{\partial t}\langle\vec{Q}\vec{Q}^T\rangle - \nabla\vec{u}^T\langle\vec{Q}\vec{Q}^T\rangle - \langle\vec{Q}\vec{Q}^T\rangle\nabla\vec{u} = \frac{4kT}{\zeta}\mathbf{I} - \frac{1}{\lambda_1}\frac{\langle\vec{Q}\vec{Q}^T\rangle}{1-\langle Q^2/Q_0^2\rangle}, \quad (3.24)$$

where again the characteristic relaxation time of the polymer is $\lambda_1 = \tau = \zeta/4H$. In order to non-dimensionalize Equation 3.24, the quantity $\langle\vec{Q}\vec{Q}^T\rangle$ is divided by the equilibrium length of the polymer $\langle Q^2\rangle_{eq}$ defined as:

$$\langle Q^2\rangle_{eq} = \frac{\frac{3kT}{H}}{1+\frac{3kT}{HQ_0^2}}. \quad (3.25)$$

This permits to redefine the conformation tensor as

$$\mathbf{C} = \frac{\langle\vec{Q}\vec{Q}^T\rangle}{\frac{1}{3}\langle Q^2\rangle_{eq}}, \quad (3.26)$$

$$C^2 = \frac{Q^2}{\frac{1}{3}\langle Q^2\rangle_{eq}}, \quad (3.27)$$

which at the equilibrium leads to $\mathbf{C} = \mathbf{I}$, leading to:

$$\frac{\partial C_{ij}}{\partial t} + u_k\frac{\partial C_{ij}}{\partial x_k} = \frac{\partial u_i}{\partial x_r}C_{rj} + C_{ir}\frac{\partial u_j}{\partial x_r} + \frac{1}{\lambda_1}\left(1+\frac{3kT}{HQ_0^2}\right)\mathbf{I} - \frac{1}{\lambda_1}\frac{C_{ij}}{1-(C^2/C_0^2)}. \quad (3.28)$$

Now the maximum distance between polymers bead can be then defined as

$$L_{max}^2 = \frac{Q_0^2}{\frac{1}{3}\langle Q^2\rangle_{eq}}, \quad (3.29)$$

so that:

$$\frac{\partial C_{ij}}{\partial t} + u_k\frac{\partial C_{ij}}{\partial x_k} = \frac{\partial u_i}{\partial x_r}C_{rj} + C_{ir}\frac{\partial u_j}{\partial x_r} - \frac{1}{\lambda_1}\left[\frac{L_{max}^2-3}{L_{max}^2-C^2}C_{ij} - \delta_{ij}\right]. \quad (3.30)$$

When normalized by $\langle Q^2 \rangle_{eq}$ the stress $\mathbf{T} = T_{ij}$becomes

$$T_{ij} = nkT[\frac{L_{max}^2 - 3}{L_{max}^2 - C^2} C_{ij} - \delta_{ij}]. \tag{3.31}$$

It is possible to write nKT as a function of the polymer viscosity knowing that

$$\eta_p = nkT\lambda_1 \frac{b}{b+3}, \tag{3.32}$$

where $b = HC_0^2/kT = L_{max}^2 - 3$. For dilute polymer solutions b is usually large enough to permit to approximate $\eta_p \approx nkT\lambda_1$ so that the polymer stress becomes [6]:

$$T_{ij} = \frac{\eta_p}{\lambda_1} \left[\frac{L_{max}^2 - 3}{L_{max}^2 - C^2} C_{ij} - \delta_{ij} \right]. \tag{3.33}$$

$$\frac{1}{2}\frac{Du^2}{Dt} = \frac{\partial}{\partial s_j}\left(u_i p \delta_{ij} + 2\nu u_i s_{ij}\right) + \frac{\partial}{\partial x_j}(u_i T_{ij}) - 2\nu s_{ij} s_{ij} - T_i j \frac{\partial u_i}{\partial x_j} + f_i u_i, \tag{3.34}$$

where the term $T_{ij}\frac{\partial u_i}{\partial x_j}$ is the rate of energy transfer to or from the potential energy stored in form of polymer stretching. This is called free energy and is given by [3]:

$$A_p = -\frac{\eta_p}{2De}\{(L_{max}^2 - 1)\log[\frac{L_{max}^2 - Tr(C_{ij})}{L_{max}^2 - 3}] + \frac{1}{3}\log(det\mathbf{C})\}. \tag{3.35}$$

The rate of variation of the free energy of the polymers is given by:

$$\frac{DA_p}{Dt} = T_{ij}\frac{\partial u_i}{\partial x_j} - \frac{1}{2De} tr(T_{ij})[\frac{L_{max}^2 - 3}{L_{max}^2 - Tr(C_{ij})}]. \tag{3.36}$$

The second term on the left-hand side of Equation 3.36 is definite positive and it hence represents the dissipation of free-energy due to the Stokes friction of the solvent on the bead [21]. This becomes an additional source of dissipation for the whole polymer-solvent system. During the coil stretch transition it reaches the same magnitude of the solvent viscous dissipation and keeps growing at higher mean extensions.

3.3.1 Limitations of FENE-P model

The FENE-P model works best for long-chain stretched polymers with no branching and a certain degree of flexibility. In such conditions, it is able to capture the main rheological behaviour of dilute polymer solution flows, among which the shear thinning, drag reduction, alteration of the energy spectra [6,41,87]. Nevertheless it fails to capture quantitative properties of the flow. Reducing the ensemble of polymer chains to one single representative dumbbell configuration, the FENE-P loses all the information on higher order moments of the configurations of the ensemble. Multi-dumbbell models and experiments on DNA molecules [39,53,101] demonstrated the existence of several relaxation times and modes while the FENE-P only represents one, which is commonly imposed to be the slowest one. This arises from the reduction of the polymer ensemble in a fluid parcel to one single representative polymer conformation as well as from the single dumbbell approximation. Indeed, depending on the assumed shape, in a multi-element chain, different numbers of elements with different extensions and orientations can react to the imposed stress, leading to different reaction forces for a given end-to-end distance and orientation. Hence, the FENE-P model tends to underestimate the reaction force when the polymer is at the equilibrium length or at a low extension [39,90]. Also due to the removal of the excluded-volume forces, the stress at low extensions is poorly represented by the model. Similarly, the assumption of a uniform Brownian bombardment on the polymer beads can lose validity in coiled sections of the polymer and prevents a number of phoretic effects on the dumbbell [50]. The kinetic theory also neglects polymer-polymer interactions, which nevertheless have been shown to take place for dilute solutions as well [49]. Additionally, excluded volume and intra-molecular interactions are responsible for the erroneous prediction of the polymer concentration effects [15,77]. Finally, properties at the atomistic levels are at large not incorporated in the kinetic theory in such a way that dependence on solvent-polymer forces, effect of the monomer size and shape on the Brownian force, effects of temperature on the equilibrium length and so on are not directly accounted for.

The difficulties previously encountered in the use of this model for quantitative analysis reside in the difficulty to model a great number of properties of the polymer molecules, the properties of its ensemble, as well as the properties of the polymer-flow interaction through a three parameters model, namely τ, η_p and L^2_{max} with the addition of only six degrees of freedom to the system. On the other side, the success of this model of this model resides in its relative simplicity. It is indeed capable to account for some atomistic behaviours of the polymer-solvent interaction without the need to recur to a stochastic approach and the simulation of large ensembles of polymer molecules for every fluid parcel of the flow. The FENE-P hence presents advantages also over only slightly more complex models, as the various multi-element FENE models, by reducing several folds the number of degrees of freedom in the system. As it will be shown in the rest of the chapter, the simulation of flows of dilute polymer solutions requires higher resolutions compared to an equivalent Newtonian flow, making the FENE-P the only viable solution for the simulation of such flows up to today. The major drawback is that the behaviour of the model must always be confirmed against qualitative observations of experimental results.

3.3.2 Numerical issues: the high Weissenberg number problem

As it can be observed Equation 3.3 lacks a diffusive term, while in reality polymer conformation has its own diffusivity, which is so small that it can be easily neglected for the purposes of the FENE-P model. The resulting set of equation is hyperbolic and does not have a stabilizing mechanism able to prevent the formation of gradients of virtually infinite steepness in the conformation tensor field [88]. The effect is comparable to the shock formation in compressible fluids, but due to the limited knowledge in polymer dynamics it is hard to determine whether it is a problem limited to the model only or whether it is actually present in real flows. The hyperbolic nature of the viable viscoelastic models plagued the field of computational rehology of polymers since its birth. According to Martien 2005 [60] äll existing numerical methods break down when the Weissenberg number exceeds a critical value", with such value being dependent on the geometry of the problem, the polymer model, the numerical scheme, the initial and the boundary conditions and mesh. For years it has been possible to obtain stable simulations only at very modest Weissenberg numbers and even in these cases often at the price of some local loss of positive definitiveness of the conformation tensor. Before proceeding in illustrating the methods, used to stabilize the numerical methods it is important to stress out that though the formation of shocks and the numerical instability that it follows are a theoretically expected behaviour of the model, this might not reflect the physical behaviour of polymers in the flow. The absence of a dissipative mechanism in the FENE-P equation permits the formation of arbitrarily small scales. Conformation tensor fluctuations at sub-Kolmogorov scale would not have the possibility to interact with larger velocity gradients being such an interaction effectively filtered out by the fluid viscosity. Incoherent fluctuations at sub-Kolmogorov scales would remain trapped in such a range and would most likely result in an increment of the molecular agitation and a shift of the equilibrium length of the polymers. There is hence the possibility for the definition of a "dissipative"mechanism in the sense of a transformation of kinetic energy from turbulent fluctuations in the non-recoverable or anyway low-grade energy form of sub-Kolmogorov scale exchange of energy between polymer stretching and molecular agitation. During the years, a number of techniques have been developed in order to deal with the high Weissenberg problem, many of these being in some measure derived from the methods used in compressible turbulence. All of them, by reducing the gradients to a manageable level, have the effect to introduce enough dissipation to the scheme in order to make it stable.

It shouldn't be surprising that one of the earliest and simplest stabilization method is the direct addition of an artificial diffusive term to the evolution of the conformation tensor in the form of $\chi\nabla^2 C_{ij}$. When enough artificial diffusion is added, it is possible to increase the stability up at higher Weissenberg and Reynolds numbers. When using this added diffusivity, it is usually considered advisable to keep its value as small as possible and its ratio ν/χ over the kinematic viscosity as large as possible. This can be considered as a Schmidt number Sc of the polymers and, in literature, values close to one or smaller are usually chosen for it [61, 68, 99]. By conjecturing that the small scales of the conformation tensor in first approximation act like passive objects, one can use the Schmidt number similarity to observe that $Sc \approx 1$ coincide to a Batchelor scale of the polymers $\lambda_b = \eta/Sc^{1/2} \approx \eta$. Hence the choice of $Sc \approx 1$ is efficient from a computational point of view, as both the flow and the conformation-tensor scales are equally resolved. It also gives an idea on the increase of resolution needed with decreasing Schmidt numbers. For example, a simulation which would be resolved for $SC = 1$ in a box of 256^3 points with $Sc = 10$ would require a box of 768^3 points. This is not a negligible issue as a code solving the Naviers-Stokes equations together with the FENE-P model requires 3 times more memory compared to the Navier-stokes solver alone and the computational costs are 5 times as high.

Other approaches try to maintain the positive definiteness of the conformation tensor through a different mathematical formulation of the problem. Fattal and Kupferman (2004) [31] for example solved the evolution of the conformation tensor in its logarithmic representation, Balci et al. 2011 [2] used a square-root-conformation representation instead, while Vaithianathan and Collins 2002 [88] applied matrix decompositions to the conformation tensor. These stabilization methods cannot completely overcome the high Weissenberg number problem [12], but they can moderately increase the range at which the computation remains stable. Only the method developed by Fattal and Kupferman allows computations at considerably high Weissenberg numbers but it cannot guarantee accuracy at these numbers.

4 Simulation of a turbulent front in dilute polymer solutions

4.1 Set-up of the numerical experiment

One of the principal objectives of this thesis is to understand the relevant mechanics to the propagation of turbulent fronts by studying such propagation in flows with altered turbulent dynamics. In order to do so, flows with polymers have been taken in considerations due to their peculiar properties. Now when one wants to study the turbulent/non-turbulent interface with a polymer model, the approach followed in Chapter 2 presents some drawbacks. First, it only allows to study a decaying flow, this prevents the computation of time averages and requires instead the use of ensemble averages that need many more independent realizations in order to converge. It also makes it hard to discern effects related to the decay of the flow from the ones related to its spatial evolution. With the introduction of the polymer model, the problem of defining a meaningful initial condition for the conformation tensor also arises. Artificially reducing the conformation tensor to the minimum extension in a similar way to the velocity damping easily leads to loss of definite positiveness, moreover it opens the question on how the arbitrarily imposed profile of conformation would affect the physical behaviour during the decay. While using non-stretched polymers at the beginning of the decay would solve the former problem, this would aggravate the latter. Finally, as it has been shown in Chapter 3 polymers can lead in both experiments and simulations to contradictory behaviors depending on the chosen parameters, initial and boundary conditions [27, 54, 67]. At the same time the FENE-P model chosen for the present study is known to give some unsatisfactory results in certain set-ups [39] and to the knowledge of the author it has not been validated before in the framework of turbulent/non-turbulent interfaces without mean shear. A validation against the dynamics observed in experiments is required and again the set-up of Chapter 2 is not suitable as it is not easily reproducible in experiments.

A different set-up has been then sought, such as to remain in the framework of turbulent flows without mean shear and one of the most diffused classes of such experiments are probably the ones utilizing oscillating grids in water tanks. The validation against such a particular reference case had been possible thanks to the availability of data from experiments performed at the Turbulence Structure Laboratory of Tel Aviv University. Despite being limited by the constraint imposed by the available technologies, these measurements are suitable to validate the macroscopic behaviour of the FENE-P model in the shearless turbulent/non-turbulent interface. The modelling approach and its validation against experiments presented in the rest of the chapter are published in Cocconi, De Angelis, Frohnapfel, Baevsky, Liberzon 2017 [16].

4.2 Modelling oscillating grid turbulence

4.2.1 Properties of oscillating grid turbulence

Performing a resolved DNS of a moving grid with the FENE-P model would be extremely computationally expensive, further reducing the maximum attainable Reynolds number of the simulation. A more efficient approach is to produce a grid forcing model stirring a region of the flow in such a way to reproduce the turbulent scales and dynamics of an actual oscillating grid. In order to do so, some of the properties of these flow must first be discussed. A test rig for the study of oscillating grid turbulence is commonly constituted by a transparent water tank with a square cross-section in which a square grid is immersed. In most experiments, the grid area completely fills the tank's horizontal cross-section trying to leave the smallest possible gap between the tank walls and the grid. Via vertical rigid rods the grid is then connected to a motor that imposes the periodic vertical motion. Thompson and Turner 1975 [82] and Hopfinger and Toly 1976 [45] performed extensive studies of the turbulence generated by oscillating grids of different size and shapes in mixing across density interfaces. They found that, with certain choices of the grid and within defined operation bounds, the turbulent flow can be characterized in terms of few a parameters. In particular when the right grid geometry is chosen, the relevant parameters are the depth of the vertical oscillation motion or stroke S, the oscillating frequency f, the distance between contiguous holes in the grid or mesh size M, the solidity ratio, i.e. the ratio between the total cross-section area of the tank and the total projected area of the grid bars. It has been found that for a grid made of square bars, within well defined range of values for the aforementioned parameters, the flow shows a number of properties that make it suitable for the study of turbulent mixing [43, 45, 64, 82].

For example, the root mean square of the velocity fluctuations decays in inverse proportion to the distance x in the inhomogeneous direction [45]. On the opposite, inertial scales as the integral length scale or the Taylor microscale λ are found to grow linearly with the same distance in such a way that both the turbulent Reynolds number $Re = l_0 u/\nu$ and Re_λ are approximatively constant within the turbulent region [100]. The reported ratio between in-plane and perpendicular velocity fluctuations are typically in the range of $1.1 - 1.2$ [45] making oscillating grid turbulence attractive for studies on quasi-isotropic turbulence.

The turbulent properties of oscillating grid flows are the result of the correct interaction between the wakes and the jets generated behind the grid bars. It should hence generally be expected a departure from the aforementioned properties in the vicinity of the grid where those jets and wakes are still in formation. Cheng and Law 2001 [13] measured the mean values of velocity fluctuations and Reynolds stresses at different grid-plan locations, showing that up to $2 - 3M$ from the grid mid-position the flow over a bar exhibits stronger rms of both u and v together with inhomogeneities in the locations over grid crossings. Regarding the outermost part of the turbulent region instead, in many of these experiments it is unfortunately unclear where the turbulent/non-turbulent interface is located, if present at all. Also, many of the experiments mentioned above were generally conducted in a steady state condition, which means that before starting the measurement the flow is stirred until an equilibrium between the energy injected and the dissipation is reached. However, a number of studies can be found in literature about non-steady properties of oscillating-grid turbulence. Most of these works had the purpose of identifying the speed of propagation or the mixing region of a turbulent flow. In a theoretical work, Long 1978 [57] first proposed a relation for the position of the interface between turbulent and non-turbulent fluid. For a flow at high

Reynolds number the average distance H of such interface from the turbulence source can be plotted as

$$H \propto (Kt)^{1/2}. \tag{4.1}$$

Here K is a parameter called "grid action", which is constant once a series of experimental parameters are kept constant. Equation 4.1 had found good agreement with experimental results [23, 24, 43], even though also other propagation rates have been reported in literature [52] under tight confinment or rotation. In any case, none of the experiments mentioned above report what happens in the final stage of propagation when the spatial decay of turbulence fluctuations reaches an equilibrium with the local dissipation, preventing the further propagation of the front.

4.2.2 Body force model implementation

Despite the popularity in the experimental community, only few attempts can be found in literature of simulations of oscillating grid. The simulation of an oscillating grid with its physical moving boundaries are too computationally expensive, so the existing simulations mostly relied on a selection of boundary conditions or body forces able to reproduce the effects of such a grid without actually simulating it. As an example, Holzner et al. 2008 [43] imposed a set of time-dependent, random velocities with length and time scales comparable to the ones of an oscillating grid. Here we follow a similar approach, but instead of directly imposing the velocities at one side of the domain we introduce a body force and instead of forcing only over the 2-dimensional boundary of the domain we force in a 3-dimensional region with finite thickness.

The body force distribution in space and time is determined by the following procedure. First a random amplitude distribution $A(y, z, t)$ in the $y-z$ directions is generated, this is done by assigning random values $\in [-1, 1]$ at equispaced nodes with separation $M = 2\pi/8$, the amplitude distribution is then obtained in the remaining points of the $y - z$ plane by a bi-cubic interpolation in space intersecting the randomly assigned nodes. A new random distribution is generated periodically with a frequency $1/\mathrm{T}_f$. The passage between two amplitudes distributions in time, $A(y, z, n\mathrm{T}_f)$ and $A(y, z, (n+1)\mathrm{T}_f)$, with $n \in \mathbb{N}$, is moreover smoothed by interpolating the two configurations in time, which produces a function $\tilde{A}(y, z, t)$. The forced region is periodic in the $y - z$ cross-section of the domain, while it remains confined to a thickness of around M in x-direction. The final 3-dimensional time varying distribution of the forcing $f(x, y, z, t)$ is given by

$$f(x, y, z, t) = \frac{K}{2}\left(1 + \tanh\left(\frac{a\Delta}{2} - a|x|\right)\right)\tilde{A}(y, z, t), \tag{4.2}$$

where the parameter K sets the intensity of the body force while Δ and a determine the thickness of the forced region. The transition in time between two different random configurations is given by the two functions $\phi_0(t)$ and $\phi_1(t)$:

$$\phi_0(t) = \frac{\Delta t - mod(t, \mathrm{T}_f)}{\mathrm{T}_f}, \tag{4.3}$$

$$\phi_1(t) = \frac{mod(t, \mathrm{T}_f)}{\mathrm{T}_f}, \tag{4.4}$$

where *mod* is the modulo function. An example of the body force distribution at a given time is depicted as a height map in figure 4.1.

Figure 4.1: 3D representation of the amplitude distribution $A(y, z, t)$ generated by the forcing model.

4.3 Validation and flow properties

As a first step, a set of Newtonian simulations have been performed in order to validate the forcing model against some known experimental results on oscillating grid turbulence. All the simulations presented hereafter were performed over a computational grid of $512 \times 256 \times 256$ Fourier modes before de-aliasing. The simulation domain is a box with dimensions $L_x \times L_y \times L_z = 4\pi \times 2\pi \times 2\pi$. The body-force modelling the grid energy input is added in a region of dimensions $\Delta \times L_y \times L_z$ at the centre of the domain as shown in figure 4.2. The

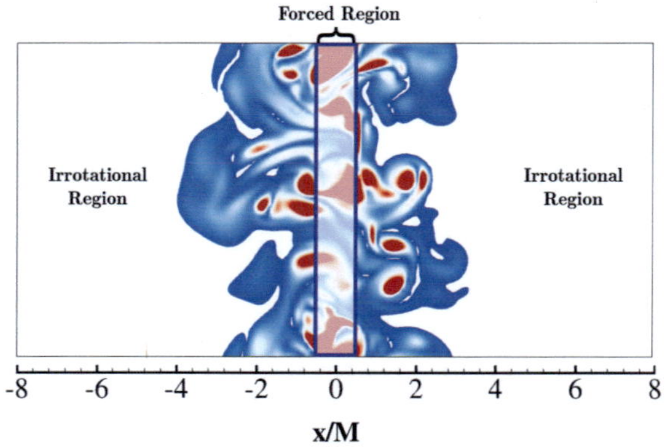

Figure 4.2: Depiction of the computational box with a slice of the enstrophy iso-contour for a Newtonian simulation. Adapted from [16].

non-dimensional viscosity is set by the choice of Re, which is in this case is $\nu = 1/Re = 1/200$, while the simulation time step is $\Delta t = 0.002$. The standard configuration for the body force is the one with 8×8 collocation points for a characteristic forcing length of $M = 2\pi/8 = 0.785$, the maximum amplitude is $K = 3.8$, $a = 1.5\pi$ and the thickness parameter is $\Delta = 0.065$. This sets the total thickness of the region where the forcing is above 10% of the maximum value to be about $\pm 0.5M$. For convenience, the bulk of the flow is defined to be the two planes at a distance of $0.6M$ from the mid-plane and these planes will be used to compute statistics relatively unperturbed by the forcing, but still close to the maximum turbulence level of the flow. For example, this permits to define an eddy turnover time as the ratio between the root mean squares of velocity fluctuations in the bulk and the length scale M. Finally, the update frequency of the forcing which determines the correlation time of it is every $T_f = 0.1$. The Newtonian statistics presented in the following chapter refer to a statistical data-set of 10 independent simulations of the duration of about 16 eddy turnover times defined as M/u'_b. All simulations start from an initial condition of quiescent fluid. When the forcing action is introduced, the fluid is perturbed and a patch of turbulence rapidly forms at the centre of the domain. The two planar-symmetric turbulent fronts then start to advance into the irrotational region of the flow. The two fronts keep propagating until the enstrophy locally advected and generated reaches an equilibrium with the one dissipated and the local enstrophy remains constant. When this happens the average position of the interface, as detected by the enstrophy threshold technique, remains constant in time. Weak vorticity, which does not possess the self-sustaining

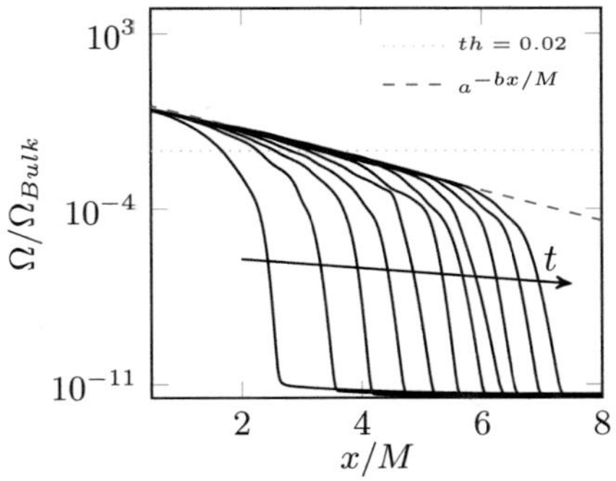

Figure 4.3: Time variation of enstrophy.

mechanisms of actual turbulent fluctuations anymore, can still diffuse into the non-turbulent region as can bee seen from figure 4.3. This also highlights the risk that a too low vorticity threshold would not be capable to capture the actual limit of the region where the flow manifests all the properties of turbulence (i.e. increased mixing, energy cascade, self-amplification of fluctuations and so on). It will be shown as other turbulent properties are required to validate the choice of a threshold value.

As can be observed from Figure 4.4, when one considers the initial growth of the turbulent patch, this follows the expected $\propto \sqrt{t}$ law. This growth slowly declines until it reaches a quasi-stationary state.

Table 4.1: Newtonian simulations parameters.

A_0	Δ	Δt	1/Re	M	Δt_u	Re_λ	η	M/u_0
3.8	0.065	0.001	0.005	$L_{y-z}/8$	0.1	50	0.037	1.92

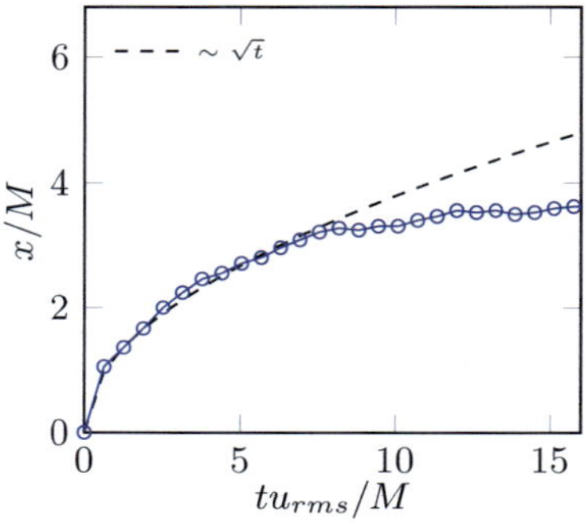

Figure 4.4: Average interface position with respect to the time.

How can be seen from Figure 4.5, the decay of the velocity fluctuations $u_{\|}$ in the in-plane direction has been observed to follow a slope proportional to $(x-x_0)^{-1}$ in the turbulent region, where $x_0 = 0.09M$ is a virtual origin slightly offset from the domain origin [45]. The fluctuations decay faster than the fitted curve after the average interface position, but this is in accordance with the observation that velocity fluctuations decay faster outside the turbulent/non-turbulent interface (specifically as a function of $(x-x_I)^{-}4$ from the local interface position x_I [19]). The integral length scale $l_0 = \int E(k)k^{-1}dk / \int E(k)dk$, depicted in Figure 4.5, as expected shows a linear increase within the turbulent region. As a further proof of the good choice of the threshold value for the interface detection, it can be noticed that the growth of l_0 departs from the observed trend of the turbulent region at the sampled average position of the turbulent/non-turbulent interface. The initial propagation of the interface also evolves according to a $\sqrt{t}$ law before slowing down and reaching a statistically steady position. In order to rule out effects of the domain size, a simulation has been performed in a domain with double the extension in x-direction and no effect on the interface propagation was observable. When looking at the anisotropy of the velocity

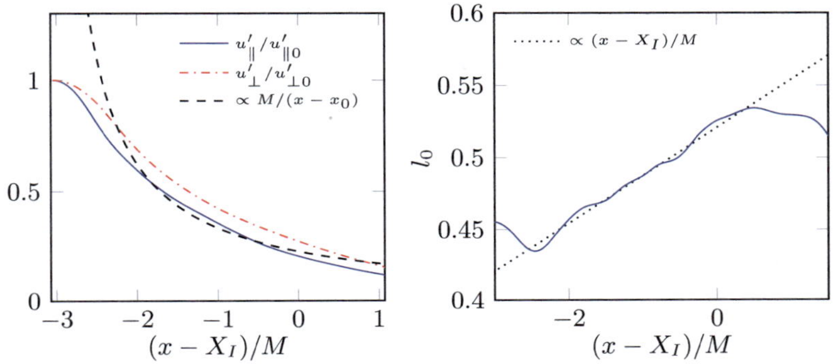

Figure 4.5: (a) Profiles of the magnitude of velocity fluctuations in the direction parallel to the homogeneous planes $y - z$ and normal to them as a function of the distance from the average interface position. The black dashed line represent the fit with a function inversely proportional to the distance from a virtual origin in the forced region. (b) Variation of the integral length scale as a function of the distance from the average interface position.

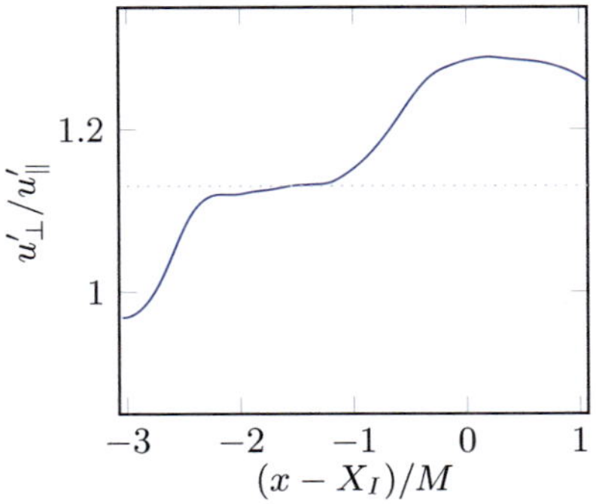

Figure 4.6: Anisotropy in the magnitude of velocity fluctuations as a function of the distance from the average interface position.

fluctuations in Figure 4.6 it is possible to see how the body force produces a nearly isotropic flow in the centre of the domain. In the bulk the ratio between the out-of-plane fluctuations $u'_{\perp}$ and the in-plane ones $u'_{\parallel}$ grows due to the inhomogeneity of the flow until a constant value of 1.13 is reached. This value is well within the range of values reported in experimental literature [45] but getting close to the interface, as expected, the anisotropy further increases until it reaches a maximum value of 1.28 at the interface.

For the viscoelastic simulations, the same set-up has been used as in the Newtonian case. One of the purposes of the present work is to discern possible local effects of the polymers at the interface from the general effect on the bulk of the flow. In order to do so a special set-up has been conceived in order to obtain a viscoelastic turbulent flow with bulk properties closer to its Newtonian counterpart. In fact, the forcing amplitude in the viscoelastic case has been tuned in order to obtain a similar steady-state integral of the energy computed among the point within the turbulent volume of the domain for the two simulations. In both cases, this should allow the interfaces to drain from a comparable supply of turbulent fluctuations, making the differences at the interface that are due to the local effect of the polymers more apparent.

The polymer diffusivity χ has been set to be equal to the kinematic viscosity such that $\chi = 1/200$, $\tau = 2$, $L_{max}^2 = 5000$ Here we focus only on phenomenological effects of the polymers on turbulent dynamics, thus despite different sets of parameters have been tested, only the results from one of these sets will be shown.

Experiments with a similar set-up have been produced by the Turbulence Structure Laboratory of Tel Aviv University. Instead of a planar grid, in their set-up a spherical agitation grid in a water tank has been used. This was justified by the necessity of avoiding any interaction between the turbulent front and the walls of the tanks while maintaining a flow with only one inhomogeneous direction (the radial one). The experimental observations will be used to verify that the FENE-P model is able to reproduce the same flow features. The Reynolds number in the experiment has been defined similarly to the simulation on the average mesh size M of the agitation device and the root mean square of the velocity fluctuations outside the envelope of movement of the grid. As for the Newtonian case, the kinetic energy in the viscoelastic simulations initially increases and reaches a stationary state after a few eddy turnover times. The main difference regards an initial overshoot in the viscoelastic simulation that precedes the steady state. It is nevertheless well known that pre-averaged models as the FENE-P perform poorly in transient flows and that an accurate prediction of the polymer stress before sufficient extensions are reached requires multi-mode models [101]. This may be imputable to a delay in the äctivationöf the polymers, as it takes some time to the random fluctuations

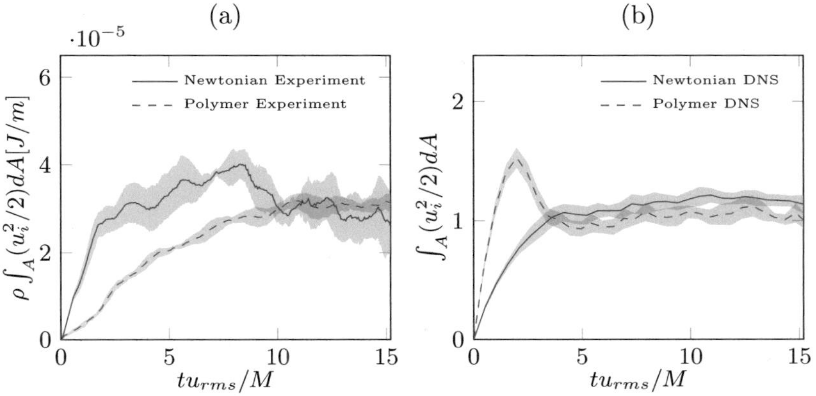

Figure 4.7: Time variation of the integral of the kinetic energy contained in the region between the bulk and the interface. Gray areas mark ±1.96 standard errors of the mean. Adapted from [16].

in the flow to stretch the polymers in an appreciable way. The forcing model on the other hand can also be responsible for this delayed polymer effect, as initially velocity gradients are not yet developed down to the smallest scales of the spectrum. Once the polymers in the bulk start to stretch significantly, the energy visibly drops and stabilizes. Such behaviour is not observable in Figure 4.7 for the experiments, where the energy grows slower in the dilute polymer solution when compared to the Newtonian case. This can be related to the fact that the physical grid immediately generates the sharp gradients required to stretch the polymers at its walls, while the forcing model requires the cascading process to develop before these are produced. This behaviour can be observed in a number of transitional statistics of viscoelastic simulations and often makes them ill-suited for transient flows [28, 39, 65, 69]. The combination of stronger forcing and Newtonian-like initial behaviour produces an initial growth of the patch which is faster than its Newtonian counterpart. Due to these reasons, it was not possible to study the transient growth of the turbulent patch for the viscoelastic case, but only its final statistically stationary state. Despite the differences in the transitory, in both simulations and experiment the final maximum size of the patch of the polymer flow is smaller than the Newtonian one, with a difference of about one mesh size M. The smaller patch size is not imputable to lower enstrophy levels in the vicinity of

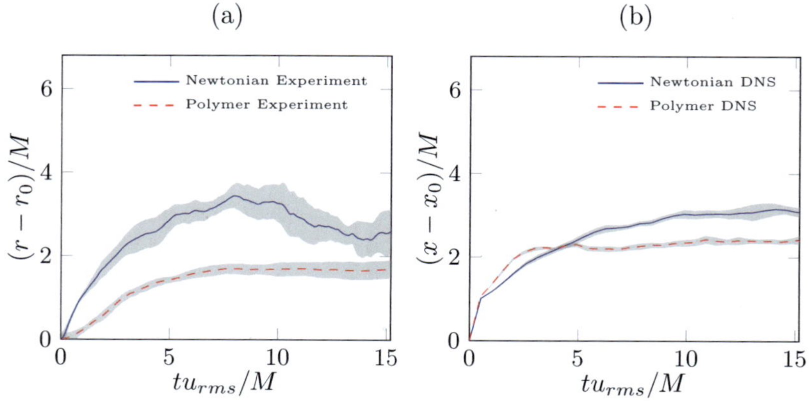

Figure 4.8: Time variation of the average interface position for Newtonian fluid and the dilute polymer solution, comparison between simulation and experimental measurements. The left $y-$axis refers to the experiments while the right one to the simulations. Grey areas mark ±1.96 standard errors of the mean. Adapted from [16].

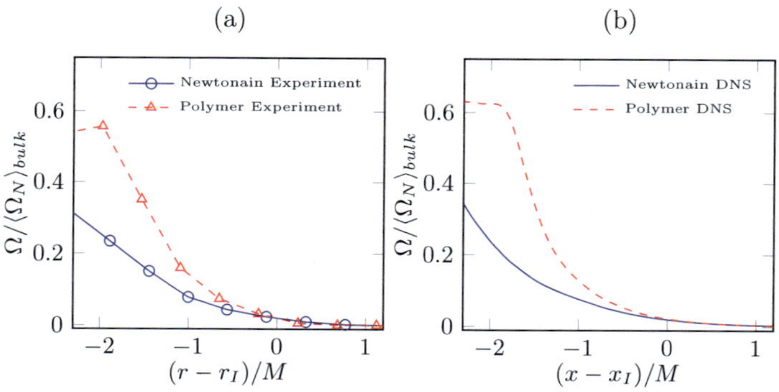

Figure 4.9: Average profiles of enstrophy as a function of the distance from the average position for the experiments (a) and for the simulations (b). The quantities have been normalized by the average enstrophy of the bulk for the Newtonian case.

the interface, as can be seen from Figure 4.9, which on the contrary is higher for the polymer case in both, experiments and simulations. A qualitative comparison of the turbulent/non-turbulent interfaces in Figure 4.10 reveals another apparent difference between the two cases: the interface produced in the polymer flow is a smoother interface with less visible small-scale features. Again, this observation has been confirmed in the experiment indicating that

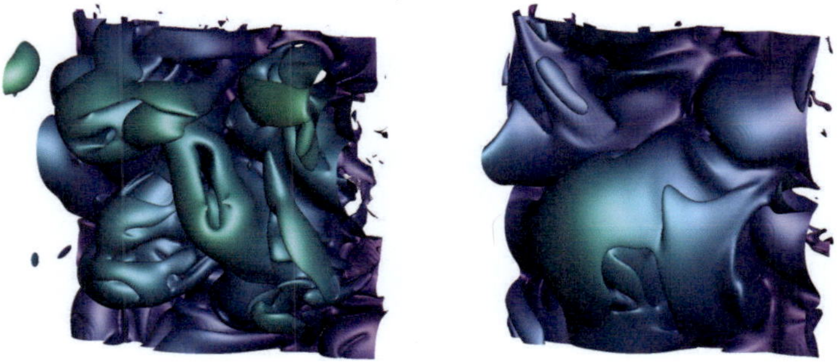

Figure 4.10: Isosurfaces of $\omega_i^2 = 0.02\omega_{i,b}^2$ for a Newtonian simulation (left) and for a simulation with polymers (right).

with the chosen set-up polymers produce interfaces that are less convoluted. To quantify reduction in convolution one can estimates the fractal dimension of the interface. Sreenivasaan et al. 1989 [76] and De Silva et al. 2013 [74], using box counting algorithms, had estimated for turublent/non-turublent interfaces a power law scaling with exponent D varying between -1.3 and -1.4 in free and wall-bounded shear flows for a fractal dimension D_f of the surface of about 2.35 ± 0.05. Following the same procedure 2D slices of the flow are divided in square sectors of equal side Δ and the number of such "boxes"containing parts of the interface is counted. The procedure is repeated for diminishing size of the boxes and the fractal dimension D is extracted from the relation $N = \Delta^D$ [74]. Figure 4.11 (a) summarizes the results of the box-counting at the steady-state in the simulations. The fractal dimension computed for the Newtonian interface obtained with a least square fit on the box count is -1.31, while when the polymers are introduced the fractal dimension drops to -1.25. Comparing the integral length scales for the two cases in Figure 4.11 (b) it appears that the polymer flow is characterized by larger scales. The integral length scale grows at a steeper rate for the polymer case producing much larger scales at the interface compared to the Newtonian fluid.

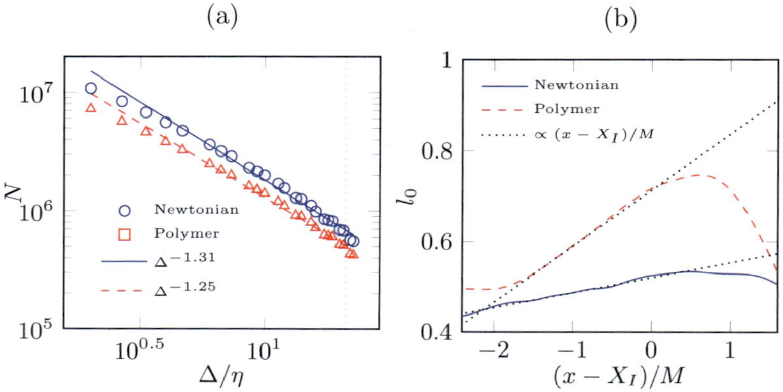

Figure 4.11: (a) Number of boxes N containing portion of the interface against box size Δ/η. The dotted line marks the box of size $\Delta = M$. (b) Integral length scale as a function of the distance from the interface.

4.4 Concluding Remarks

The forcing model developed for the simulation has been proven capable to reproduce features of oscillating grid turbulence as the quasi-isotropy of velocity fluctuations, their rate of decay, linear growth of integral scales and initial growth rate of the turbulent region. The FENE-P model, of which the application in turbulent/non-turbulent interface is scarce if not absent, has been validated against ad-hoc experiments. Agreement between simulations and experiment could not be found for the initial growth phase, though further investigations are required in order to known weather shortcomings in the forcing or in the polymer model are to be imputed. Once the steady state was reached, the FENE-P proved to reproduce a number of flow features observed in the real dilute polymer solutions of the experiments, proving yet another time the capability of this model to capture the qualitative behaviour of turbulent flows of dilute polymer solutions.

5 Results and discussion

In the previous chapter the numerical set-up has been introduced and the simulations have been compared to experiments in a similar set-up. The most evident effects of the polymers in the flow can be resumed in a reduced maximum size of the turbulent patch and a reduction in the fractal complexity of the turbulent/non-turbulent interface. In the following chapter possible reasons for these phenomena are investigated. The flow will be first investigated in the Eulearian frame of reference. In particular taking advantage of the numerical approach the interaction between polymers, vorticity and strain will be directly investigated. Comparison with the Newtonian case will be as before used in order to spot the most relevant changes introduced by polymers. At the same time statistics at the interface will be compared with the ones in the bulk of the flow, allowing to detect deviations of polymers behaviour from the ones expected for homogeneous quasi-isotropic turbulence. Statistics of orientations between vorticity field, strain eigenframe and polymer conformation tensor are also analysed in this chapter. Such statistics have been in part previously published in Cocconi, De Angelis, Frohnapfel, Baevsky, Liberzon 2017 [16]. In a second moment turbulent statistics along Lagrangian trajectories across the turbulent/non-turbulent interface will be analysed. This allows for a better assessment of the contributes on the observed Eulerian statistics at the interface from the antagonists process of entrainment of non-turbulent fluid and of the return to the laminar state of previously turbulent parts of the fluid.

5.1 Eulerian statistics

The statistics presented in this section are ensemble averaged across all the realizations performed and averaged in time in the steady state interval $tu'_b/M \in [7, 16]$ for a total of 60 independent samples. As it has been observed from

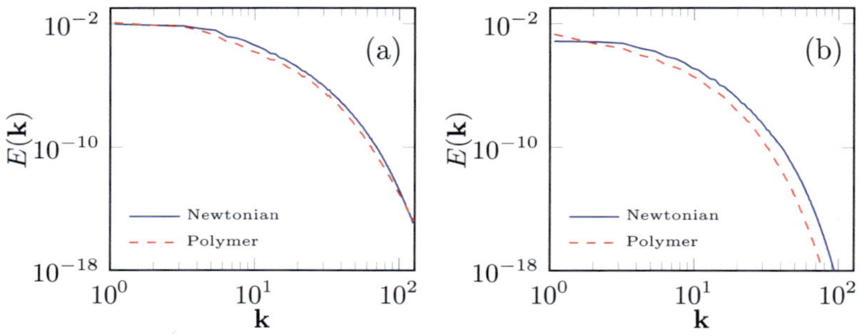

Figure 5.1: Spectra of kinetic energy compute over homogeneous planes $y - z$ in the bulk (a) and at the average interface position (b).

the integral scales, the flow with polymers lead to an increased size of the large scales. As expected also the energy spectra taken in the bulk of the flow shown in Figure 5.1 (a) confirm an increase in the energy content of the largest scales (small wave-numbers **k**) with a reduction in the energy of the smaller ones. Nevertheless, the differences are little and the two flows experience relatively similar distribution of energy in the bulk of the flow. The reduction of energy is more evident for the spectra sampled at the average interface position of Figure 5.1 (b). Indeed, looking at Figure 5.2 it can be seen how getting closer to the interface the difference in the Kolmogorov scales between the two cases increases so that both small and large scales have increased sizes at the interface for the polymer case. One of the main advantages of the numerical approach over the experimental one is the possibility to measure the state of the conformation tensor for every point of the field. In such a way orientation and extension state of the polymers can be obtained, allowing to directly study the polymer-fluid interaction. For example the average end-to-end extension of the polymers is given by the trace of the conformation tensor $tr(\mathbf{C})$. In Figure 5.3 this has been normalised by the maximum extension L_{max}^2 to show the distribution of the average stretching with the distance from the interface. One can see there how the polymers are most stretched in the forced region and the bulk, where the strongest velocity gradient resides. There average extensions up to 30% of L_{max}^2 are reached while getting close to the non-turbulent region the average extension quickly decays as the intensity of turbulent fluctuations also decay. Noteworthy is the presence of residual extension into the non-turbulent region: this can in part be explained by the

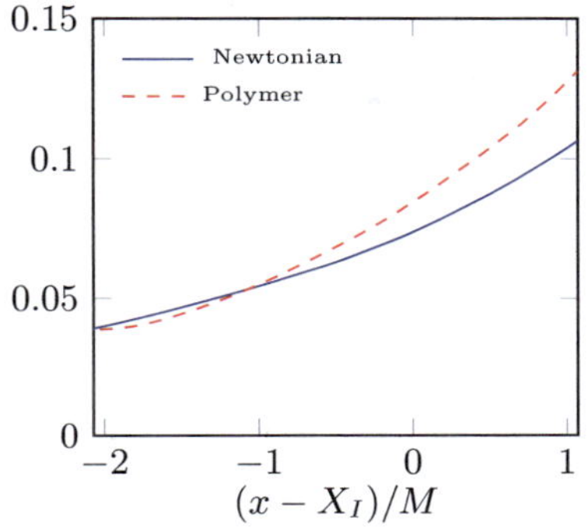

Figure 5.2: Kolmogorov length scale $\eta = (\nu^3/\epsilon)^{\frac{1}{4}}$ as a function of the distance from the interface.

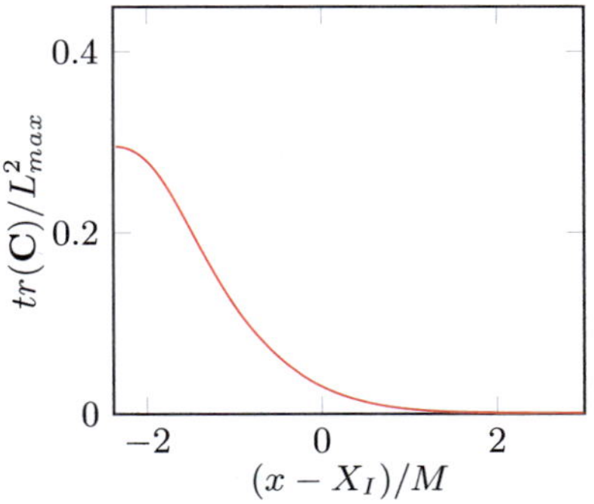

Figure 5.3: Average trace of the conformation tensor normalized by the maximum allowed extension L^2_{max} as a function of the distance from the average interface position.

presence of weak irrotational fluctuations that can still stretch the polymers and in part by the presence of fluid particles ëxtrained"from the turbulent flow that retains some degree of polymer extension. While stretching and relaxing polymers store, release and dissipate energy. Figure 5.4 (a) depicts the free energy A_p stored into polymers compared to the kinetic energy. While both Newtonian and polymer flows present comparable maximum kinetic energies, the total energy available to the flow is roughly 30% higher. Part of the excess of free energy is continuously exchanged with flow, part of it though is being dissipated by means of Stokes friction during the relaxation of the chain. This

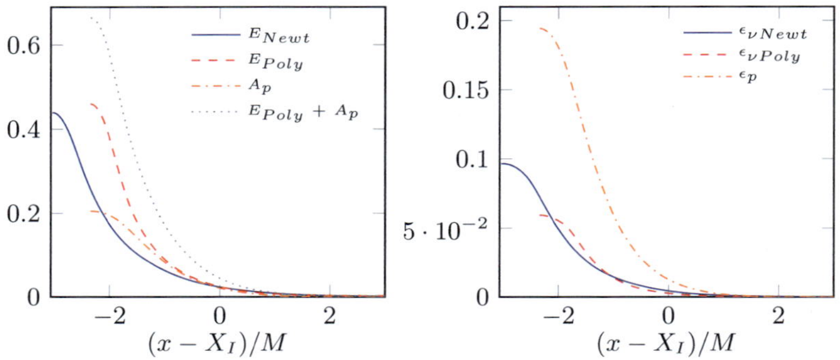

Figure 5.4: (a) Average profiles of kinetic energy and free energy of the polymers. (b) Viscous and polymer dissipation rates. Both as a function of the distance from the average interface position.

is depicted in Figure 5.4 (b) together with the viscous dissipation for the two flows. As expected, the maximum viscous dissipation rate is lower for the polymer case, while the dissipation of free-energy due to polymer relaxation results comparatively much larger. The larger polymer dissipation is ultimately responsible for the dissipation of the additional kinetic energy injected by the body force into the flow and captured by polymers and stored in form of end-to-end extension. As velocity gradients are the driving mechanism for both the unfolding and relaxing of polymers, free energy and polymer dissipation are deeply interconnected to strain and vorticity. Indeed, when the polymers are taken in account the budget equation for the enstrophy presents an extra term due to the polymer stress interacting with vorticity. When also the forcing contribution is added, Equation 2.1 becomes:

$$\underbrace{\frac{1}{2}\frac{\partial \omega^2}{\partial t}}_{\partial t} = \underbrace{-\frac{1}{2}u_j\frac{\partial \omega^2}{\partial x_j}}_{T} \underbrace{+\omega_i\omega_j s_{ij}}_{P} \underbrace{-\nu\frac{\partial \omega_i}{\partial x_j}\frac{\partial \omega_i}{\partial x_j}}_{\epsilon} \underbrace{+\frac{\nu}{2}\frac{\partial \omega^2}{\partial x_j \partial x_j}}_{D}$$
$$\underbrace{+\nu\omega_i\epsilon_{ijk}\frac{\partial^2 T_{kl}}{\partial x_j \partial x_l}}_{Ve} \underbrace{+\omega_i\epsilon_{ijk}\frac{\partial f_k}{\partial x_j}}_{f_\omega} \quad (5.1)$$

Figures 5.5 and 5.6 depicts the terms of Equation 5.1 as a function of the distance from the mean interface position X_I. The time variation has been

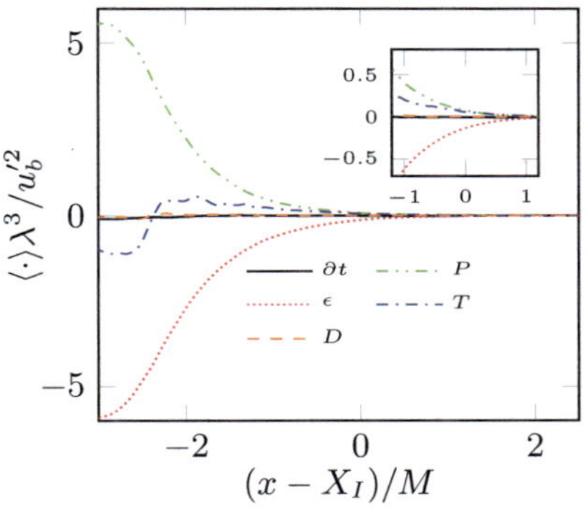

Figure 5.5: Enstrophy budget for the Newtonian case. The right scale refers to f_m, i.e. the maximum possible amplitude of the body force.

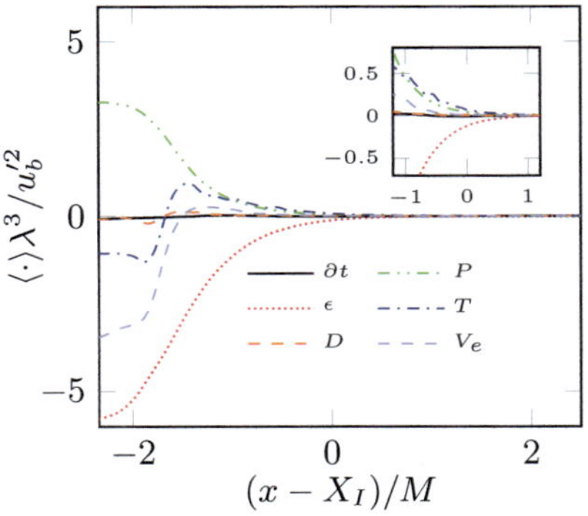

Figure 5.6: Enstrophy budget for the polymer case. The right scale refers to f_m, i.e. the maximum possible amplitude of the body force.

directly computed by finite differentiation in time of the enstrophy profiles. Both plots show how enstrophy is steady in time everywhere, even at the mean interface position there is no appreciable variation in time of enstrophy.

While looking at Figure 5.5 a parallel can be drawn with the budget studied in Chapter 2: similarly to the decaying case, the advection here draws enstrophy from the most turbulent part of the flow and releases it towards the interface. Also, similarly to the decaying case, viscous diffusion is practically negligible everywhere. For the Newtonian case in the forced region we observe the expected approximate Tenneks and Lumley balance between enstrophy production P and destruction ϵ. The production though is slightly weaker than the dissipation and the enstrophy lost through the latter and through advective fluxes is compensated by the one injected by the body-force (not shown). Getting closer to the interface the production decays faster than the dissipation and the rate of variation of enstrophy is kept constant only by the increasing relative weight of the advection. By looking at the budget for the polymer case of Figure 5.6 it can be seen how the viscous dissipation in the bulk is in magnitude about one third stronger than the production. The enstrophy sink in the forced region is further reinforced by the advection and the viscoelastic contribution V_e, this stronger sink is compensated by a much stronger enstrophy injection from the forcing compared to the Newtonian case. Particularly interesting is the behaviour of the viscoelastic term while where the turbulence is most intense it leads to a destruction of enstrophy, away from the bulk the relaxation of the polymers with its energy release leads to a small but positive contribute to the enstrophy variation. In the inset of Figure 5.6 it is also possible to see that the contribute from the polymers becomes negligible about half mesh size M before the average interface position where the enstrophy variation appears where the dominant terms are again dissipation, production and advection. Interesting is the fact that the advection appears to be more intense in the polymer flow when scaled in inertial units. It is possible that in certain condition the increased advection could tip-off the other propagation-reducing effects of polymers, thus explaining these cases reported in literature of increased propagation speed in dilute polymer solutions [54].

5.1.1 Conditional statistics

The statistics previously presented average together samples homogeneous in their x coordinate. In this way, though, the sample in the region close to the average position of the interface includes both, turbulent and non-turbulent samples. The homogeneity of the turbulent property of the sample close to the interface region can be improved by means of conditional sampling. In in interface studies this generally means that samples are averaged together when they possess the same distance from the instantaneous local distance of the interface. Details on how this distance can be defined and how conditional statistics are performed can be found in Appendix. Obviously, these statistics maintain a certain sensitivity to the arbitrary choice of the enstrophy threshold. The initial choice of a 2‰ enstrophy threshold proved to be good enough to detect the changes in conventional statistics, marking the passage between turbulent and non-turbulent region and hence constitute the reference value for the present study. Nevertheless, also statistics at 0.5‰ and 10‰ threshold will be occasionally shown in order to assess the effect of different thresholds on the robustness of the observations made. One typical observation in con-

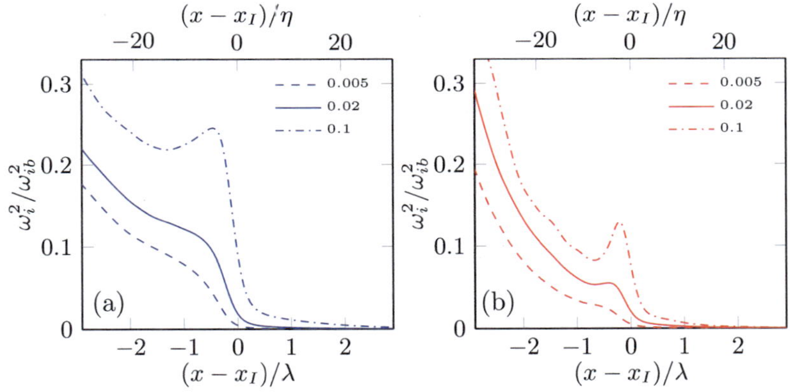

Figure 5.7: Conditional enstrophy profiles normalised by the average enstrophy in the bulk for three different thresholds $\omega^2_{i,th} = 0.1 - 0.02 - 0.005\omega^2_{i,b}$. (a) Newtonian case, (b) polymers.

ditional statistics across turbulent/non-turbulent interfaces is a steep jump of enstrophy. Here when the threshold $\omega^2_{i,th}$ is set to $0.005\omega^2_{i,b}$ (or 0.5‰ of the mean bulk enstrophy) the enstrophy jump across the interface is barely visible for the polymer case. High thresholds, as in the case of $\omega^2_{i,th} = 0.1\omega^2_{i,b}$, show a steep jump in the enstrophy profile but also introduce an unrealistic peak at the interface. The 2‰ threshold capture the jump of enstrophy across the interface for the polymer case without introducing any peak in the profile, confirming to be a reasonable threshold choice. The relatively steep jump observed in the conditional enstrophy is not observable in the conditional profiles of turbulent kinetic energy of Figure 5.8. It is interesting to observe how as the initial build-

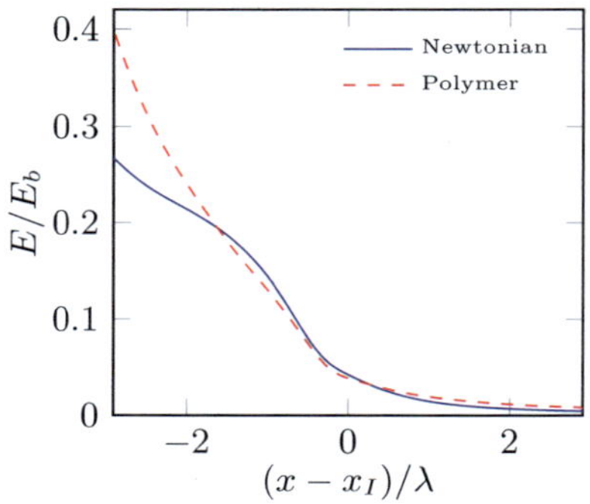

Figure 5.8: Conditional averages of kinetic energy normalised by their values in the bulk.

up of turbulent kinetic energy moving from the interface towards the bulk is comparable for the two cases. From the conditional averages it appears that the polymer flow in the interface region experiences reduced turbulent enstrophy but with very similar levels of kinetic energy. In homogeneous isotropic turbulence on average strain and enstrophy fluctuations are in equilibrium being $s_{ij}s_{ij} = \omega_i^2/2$. In Figure 5.9 for both, polymer and Newtonian flows the expected equilibrium is found in the turbulent region far from the interface. The equilibrium is nevertheless lost in the region around the interface and this is particularly apparent in the Newtonian case where, moving towards the non-turbulent region, vorticity fluctuations decays much faster than the strain rate. For the polymer case, the difference is less intense and strain rate fluctuations decrease only marginally slower than the vorticity ones into the non-turbulent region. Borrell and Jimenz [8] exploited the existence of a non-equilibrium region in their definition of interface layer for a turbulent boundary layer identifying it with it. Also, as found by Borrell and Jimenez [8] to rectilinear distances, it is possible to observe that the interface thickness for both cases scales with λ, but the polymer one appears to be sharper due to a reduction of the thickness of the non-equilibrium region, as it can be seen by the ratios $\omega_i^2/(2s_{ij}s_{ij}$ of Figure 5.10. Indeed, the region of stronger inhomogeneity has a thickness on the order of 1.4λ or about 30η for the Newtonian case while for the Polymer case we have a thickness around 0.4λ or 10η. Similarly to the strain, the conditional averages of the polymer extension of Figure 5.11 present non-zero values well into the non-turbulent region. Also, the steepness of the growth of the average extension moving towards the turbulent region is not

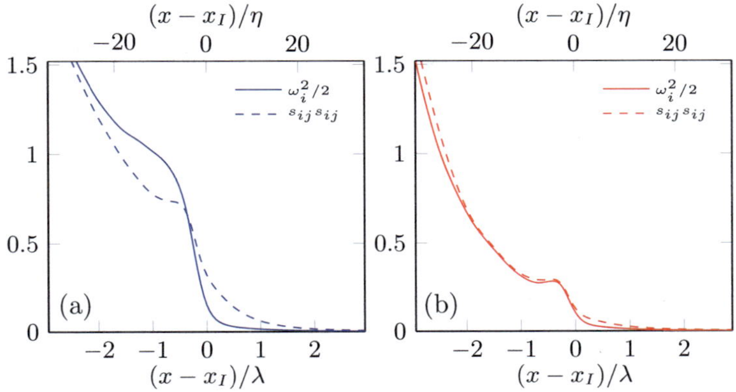

Figure 5.9: Conditional averages of enstrophy compared to the strain rate. (a) Newtonian case, (b) polymers for $\omega_{i,th}^2 = 0.02$.

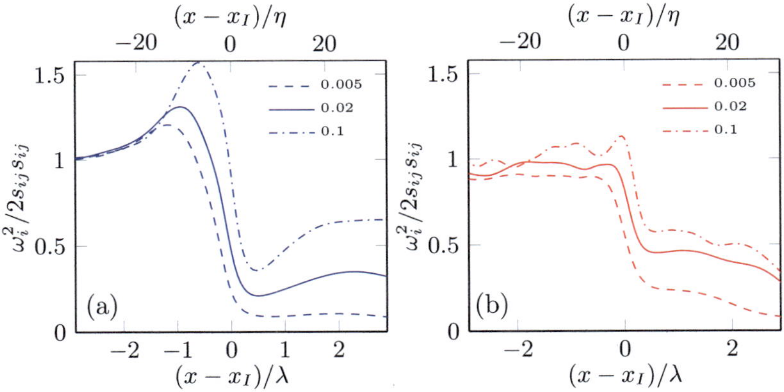

Figure 5.10: Condtional averages of the ratio $\omega_i^2/2s_{ij}s_{ij}$ for three different thresholds $\omega_{i,th}^2 = 0.1 - 0.02 - 0.005\omega_{i,b}^2$. (a) Newtonian case, (b) polymers.

as marked as for strain and enstrophy. More information can be obtained by watching at the joint *pdf* of the polymer extension respect to the distance from the interface. Figure 5.12 shows how close to the bulk of the flow the occurrences of non-stretched polymers is relatively low, while in the non-turbulent region almost all the samples have the minimum extension. It also shows that

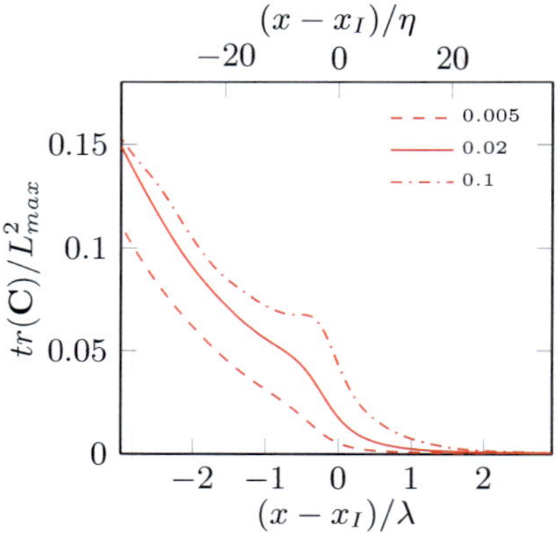

Figure 5.11: Conditional average of the $tr(\mathbf{C})$ normalised by the maximum allowed extension L^2_{max}.

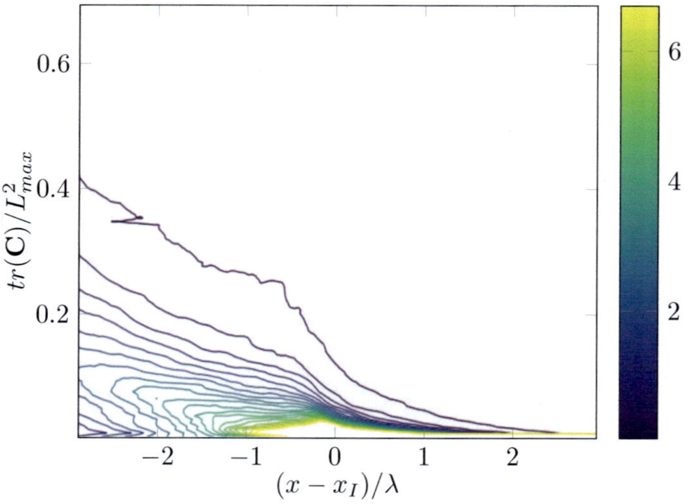

Figure 5.12: Conditional joint probability density function of $tr(\mathbf{C})/L^2_{max}$ for $\omega^2_{i,th} = 0.02\omega^2_{i,b}$

two concurrent behaviours are observable: medium to large extensions in the bulk ($0.17 < tr(\mathbf{C})/L^2_{max} < 0.4$) decay almost linearly with the distance from the interface up to a distance of about 0.3λ before the interface. On the other side in the region between -0.3λ and the interface an increasing frequency of weak extensions (around $0.03L_{max}^2$ is observable. Especially the relatively large extensions in the irrotational region are likely to be due to residual extension from extrained fluid particles that do not reach the minimum extension up to distances of about 3λ from the interface. Conversely it is likely that the high frequency of lightly extended polymers inside the turbulent region within $1 - 2\lambda$ from the interface are partially due the presence of recently entrained fluid particles. In this region a large number of fluid particles are little stretched or not stretched at all. Together the two trends produce a peak of extensions below 7% of the maximum allowed close to the turbulent/non-turbulent interface. Firstly, the peak highlights how at the steady state the two phenomena of entrainment and extrainment bring different contributes to the statistics of the interface, and secondly how the polymer action at the interface is likely to be dominated by stretching and relaxing at very low extensions. Low-extension range is known to present some difficulties for FENE models, as these tend to underestimate the polymer reaction force, overestimating its extension [39], which helps to explain the overestimation of the maximum patch size for polymers compared to the experiment. Concluding from conditional statistics it appears that polymers in the region around the interface are mostly transitioning to a quasi un-stretched state or first uncoiling from it. In the same region the ratio $\omega_i^2/2s_{ij}s_{ij}$ appears much different from the one observed for the Newtonian flow, possibly highlighting a local polymer effect on the dynamics between strain and vorticity

5.1.2 Strain and vorticity

In Chapter 2 the importance of the dynamics between strain and vorticity in the framework of turbulence propagation have been illustrated. Here it is shown how the interaction between polymers and velocity gradients alter those dynamics in both the bulk and in the turbulent/non-turbulent interface. In the FENE-P model the polymers are influenced by the velocity gradients through the terms $\frac{\partial u_i}{\partial x_r}C_{rj} + C_{ir}\frac{\partial u_j}{\partial x_r}$. This can be rewritten as $s_{ir}C_{rj} + C_{ir}s_{rj}$ [87], highlighting the direct dependence to the strain field. In the evolution of the conformation tensor the strain contribution must counteract the one from the elastic reaction force in order to stretch the polymers. It is possible to

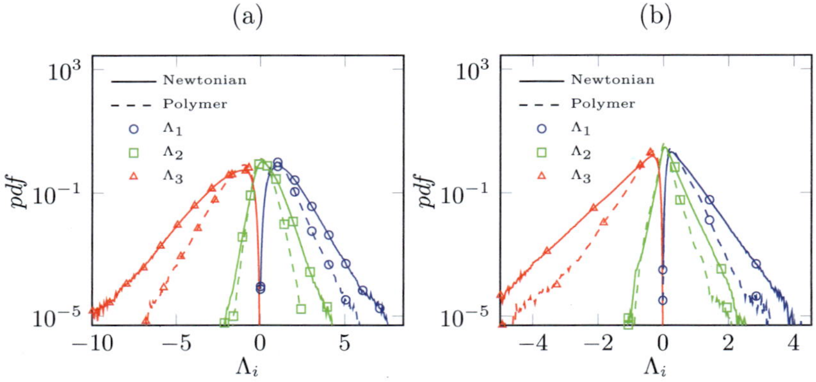

Figure 5.13: Probability density function of the strain eigenvalues Λ_i (a) in the bulk and (b) at the interface. Adapted from [16].

observe how polymers in return alter the strain field, especially considering the influence of the latter on the enstrophy dynamics. In Figure 5.13 changes in

Table 5.1: Average values of the strain rate eigenvalues from the DNS.

	Bulk		Interface	
	Newtonian	Polymer	Newtonian	Polymer
$\langle\Lambda_1\rangle$	1.34	1.12	0.49	0.39
$\langle\Lambda_2\rangle$	0.29	0.15	0.14	0.07
$\langle\Lambda_3\rangle$	-1.70	-1.28	-0.64	-0.47

the strain eigenframe are investigated through the probability density functions of its eigenvalues. In in Figure 5.13 (a) it can be seen how both polymer and Newtonian flow in the bulk show a typical relative distribution of the eigenvalues Λ_i for turbulent flows. The polymer flows apparently experience lower probability of extreme events with reduced tails of the distribution for all the eigenvalues but for the negative tails of the intermediate eigenvalue Λ_2. The same trend is observable in Figure 5.13 (b) for the interface: again, the negative events of Λ_2 have the same frequency for both the polymer and the Newtonian flow, all the other extreme events are reduced for the polymer case compared to the Newtonian one. The trend is also confirmed by the average values of Λ_i recapitulated in Table5.1 where one can also note as the largest variation between Newtonian and polymer case happens for the intermediate eigenvalue Λ_2. The main difference from the bulk is an increased weight of the tails for the polymer eigenvalues denoting that relatively extreme straining events are more frequent for the polymer case. On average $\Lambda_1\Lambda_2\Lambda_3 = -\frac{1}{4}\omega_i\omega_j s_{ij}$ so it can be expected that the reduced frequency of positive events of Λ_2 leads to a reduced positive contribution to the enstrophy production.

However, Lüthi et al. 2005 [58] demonstrated how the contribution to the enstrophy production depends more on the orientation of those eigenvector with vorticity than on their magnitude alone, as the relation $\omega_i\omega_j s_{ij} = \omega_i^2\Lambda_i cos^2(\vec{\omega}\cdot\vec{\lambda}_i)$ highlights. Thus, independently from the intensity of the vorticity and the eigenvalue Λ_i, if the associated eigenvector $\vec{\lambda}_i$ is perpendicular to the local vorticity vector, there is no net change in enstrophy. In Figure 5.14 (a) and (b)

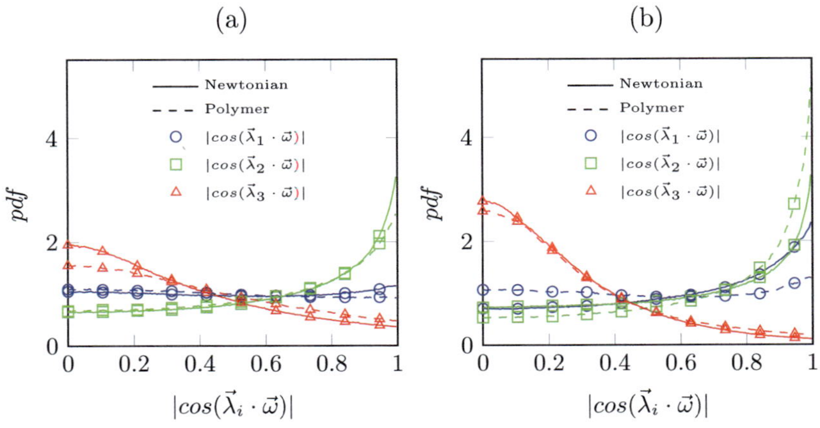

Figure 5.14: Probability density function of the cosine of the angle between the vorticity vector and the three strain eigenvector $\vec{\lambda}_i$. (a) in the bulk, (b) at the interface. Adapted from [16].

the probability density function of the alignment between the eigenvectors $\vec{\lambda}_i$ and the vorticity $\vec{\omega}$ is shown respectively for the bulk and the interface. The bulk does not show any strong deviation from the Newtonian case for the polymer flow and in general it follows the same behaviour observed for a number of other turbulent flows [87]. For both it can be seen how $|cos(\vec{\omega}\cdot\vec{\lambda}_2)|$ having a higher frequency of values close to 1 shows that $\vec{\lambda}_2$ is the eigenvector most strongly aligned with the vorticity. The orientation with $\vec{\lambda}_1$ has almost equal probability for all the angles denoting a random alignment. Finally higher probability of values of the cosine close to zero means that $\vec{\lambda}_3$ is more likely found to be orthogonal to the vorticity vector. The only notable difference between polymer and Newtonian flow in the bulk is indeed a higher probability that $\vec{\lambda}_3$ and vorticity are orthogonal for the Newtonian case. In Figure 5.14 (b) the same alignments are depicted for the turbulent/non-turbulent interface. It is useful to remember that in this region the vorticity has to align with the interface and hence orientation between strain and vorticity is also representative of the orientations between strain and the interface itself. Comparing the Newtonian case at the interface with the bulk an increased alignment with the $\vec{\lambda}_1$ is observable, while the alignment with the compressive eigenvector $\vec{\lambda}_3$ is further reduced. Henceforth for the Newtonian flow an enhanced enstrophy production can be expected at the interface due to a combination of reduced negative contribution from compressive events ($\vec{\lambda}_3$) and increased positive one from stretching events($\vec{\lambda}_1$). A similar trend is observed for $\vec{\lambda}_3$ in polymers, but in this case a further increase in $\vec{\lambda}_2$ alignment is observed without the increase in $\vec{\lambda}_1$ alignment observed for the Newtonian case. Lüthi et al. [58] noted how the strongest positive contributes to $\omega_i \omega_j s_{ij}$ comes from $\omega_i^2 \Lambda_i cos^2(\vec{\omega} \cdot \vec{\lambda}_1)$, hence from the vorticity being parallel to $\vec{\lambda}_1$. Therefore, a stronger $\vec{\lambda}_2$ alignment leads to weaker, yet positive, contributes to the enstrophy production due to the fact that the eigenvalue Λ_2 can have negative values and its positive values are smaller compared to Λ_1 ones. In Figure 5.13 (b) it has been seen that the *pdf* of Λ_2 in the polymer case, compared to the Newtonian case, shows a reduced frequency of positive events without a comparable reduced frequency of negative ones. This fact, coupled with the observed alignment, further moves the balance in favour of negative enstrophy production events for the polymer flow at the interface. The change in those in alignments is connected to the interaction between polymers and the velocity derivatives field so it is interesting to study how these stretch with respect to the strain eigenframe and vorticity. Similarly to the strain rate the polymer orientation and extension state in the FENE-P model has the representation of a tensor, its directionality is determined by its eigenframe and in particular the polymer orientation is represented by the eigenvector associated to the strongest eigenvalue of the conformation tensor. Figure 5.14 (c) depicts the cosine of the angle between the principal polymer eigenvector $\vec{\epsilon}_1$ and the strain eigenframe $\vec{\lambda}_i$ in the bulk and at the interface. For the interface only those samples with a minimum maximum eigenvalue of 1.5 have been used in order to avoid the contribution from non-stretched polymers for which the orientation is ill-defined in the model. Figure 5.15 depicts for the bulk a picture observed before in other

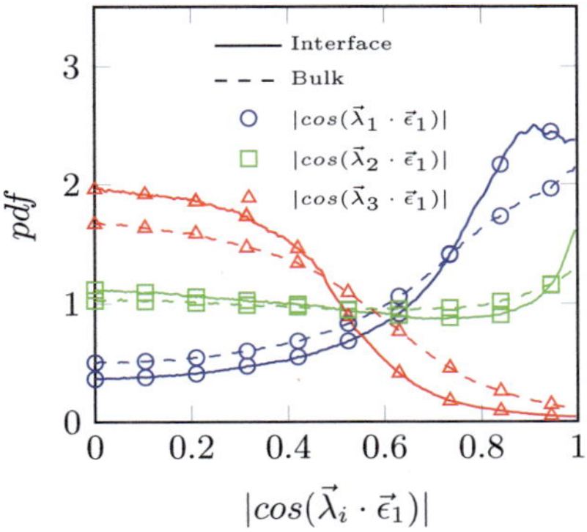

Figure 5.15: Probability density function of the cosine of the angle between the largest conformation tensor eigenvector $\vec{\epsilon}_1$ and the three strain eigenvector $\vec{\lambda}_i$. (a) in the bulk, (b) at the interface. Adapted from [16].

homogeneous isotropic turbulence studies with polymers [89,90]. In particular $\vec{\epsilon}_1$ is preferentially aligned with $\vec{\lambda}_1$, it is randomly oriented respect $\vec{\lambda}_2$ and it is more likely to be orthogonal to $\vec{\lambda}_3$. At the interface the alignment between polymers and both $\vec{\lambda}_1$ and $\vec{\lambda}_2$, is further increased as well as the orthogonality with $\vec{\lambda}_3$ denoting that polymers are more uniformly aligned with the strain eigenframe. Polymers in homogeneous isotropic turbulence are also known to be preferably aligned with vorticity [89,90]. Indeed, 5.16 shows how polymers are preferably oriented along the vorticity vector in the bulk of the flow, so that the orientation between strain and polymers partially reflects the one with vorticity. The interface sees this trend exacerbated with polymers on average almost parallel to the vorticity. It is important to note how the vorticity vector at the interface is parallel to the interface itself so that polymers can be considered strongly aligned with the latter. The preferential orientation along the interface also lets infer that, while stretching, polymers tend to oppose the inertial mechanism that try to further increase the surface of the interface.

Being aligned with vorticity, polymers can only weakly directly interact with it, in fact this would require to affect the asymmetric part of the velocity gradient

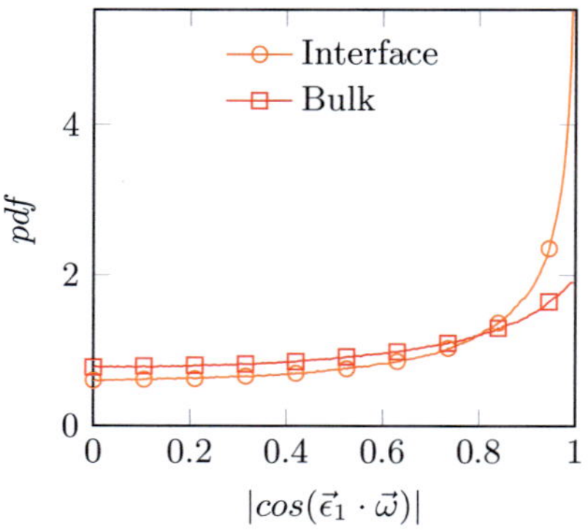

Figure 5.16: Probability density function of the cosine of the angle between the largest conformation tensor eigenvector $\vec{\epsilon}_1$ and the vorticity vector. Adapted from [16].

tensor in the plane orthogonal to the polymers. Hence, by stretching and relaxing, the polymers mainly influence the strain. This is particularly important for the vortex stretching, as when aligned to vorticity, polymers can only either adverse the stretching of vorticity or enhance vortex compression. The stronger alignments of the polymers with $\vec{\lambda}_1$ and $\vec{\lambda}_2$ observed in Figure 5.15 can be thus expected to lead to reduced enstrophy production via a combination of reduction of stretching and enhanced vortex compression. Figure 5.17 compares the

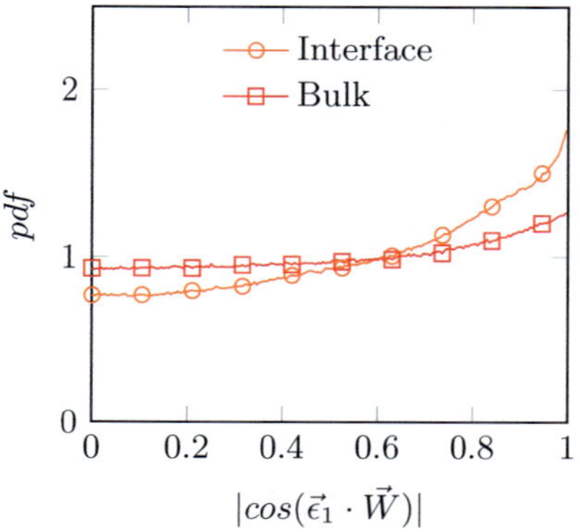

Figure 5.17: Probability density function of the cosine of the angle between the largest conformation tensor eigenvector $\vec{\epsilon}_1$ and the vortex stretching vector $\vec{\mathbf{W}} = \{\omega_j s_{ij}\}$.

orientation between the vortex stretching vector $\vec{W} = \omega_i s_{ij}$ and the polymer eigenvector $\vec{\epsilon}_1$. The plot indeed shows how the increase of alignment between polymers and the two eigenvectors $\vec{\lambda}_1$ and $\vec{\lambda}_2$ previously observed coincide with an increased alignment between polymers and the vortex stretching vector $\vec{W} = \omega_i s_{ij}$, when compared to the bulk of the flow. The final effect of these

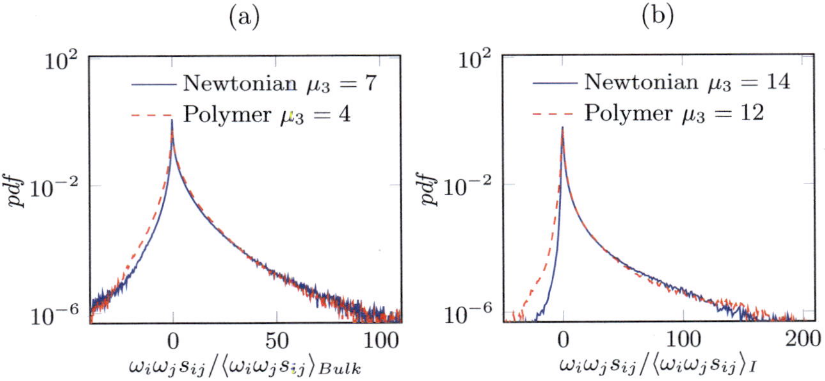

Figure 5.18: Probability density function of the enstrophy production $\omega_i \omega_j s_{ij}$ normalized by its average value. (a) in the bulk (b) at the interface, μ_3 is the estimated skewness of the distribution. Adapted from [16].

changes in the orientation statistics at the interface is to move the typical distribution between positive enstrophy production events (stretching) and negative ones (compression). In Figure 5.18 (a) and (b) such changes are investigated through the *pdf* of the enstrophy production, for both cases, in the bulk and at the interface. These distributions have been normalized by the average of $\omega_i \omega_j s_{ij}$ of the sections to which they refer in order to highlight the differences in the balance. In Figure 5.18 (a) it can be seen how the distributions for the polymer case and the Newtonian one almost collapse in the bulk. Only a small increase in weak negative events is observable for the polymer case and it appears that despite the action of the polymers that effectively introduce new stresses into the flow in the bulk, vorticity and strain reorganize according to the observed üniversal"pattern. The same does not hold anymore for the interface, as can be seen from Figure 5.18 (b). First it can be observed how for both cases the distribution between compression and stretching is strongly shifted towards stretching events when compared to the bulk. Strong stretching events can be up to eight times stronger than the strong compressive events of the same likelihood, while at the bulk they were at best two time stronger. The prevalence of stretching events at the interface is peculiar when it is observed that the interface is an inherently viscous region and viscosity in homogeneous isotropic turbulence is known to be the main bounding factor against the growth of the vortex stretching [58]. In the *pdf*s for the polymer case the significant changes in the interactions between strain and vorticity previously observed lead to an increased weight of vortex compression compared to the Newtonian one. The observed shift also appears to be robust to the choice of the threshold, as shown in Figure 5.19, and it becomes more evident when the threshold is lowered. The *pdf* of the strain rate production $s_{ij}s_{jk}s_{ki}$, unlike

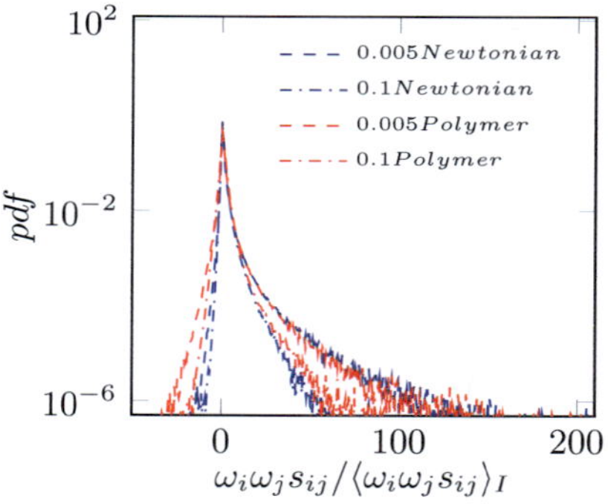

Figure 5.19: Probability density function of the normalized enstrophy production $\omega_i\omega_j s_{ij}$ at the interface with different thresholds $\omega^2_{i,th}/2$ for the interface sampling.

$\omega_i\omega_j s_{ij}$, shows how polymers shift towards negative events in both the bulk and the interface. How it has been noted before, the strain field is more directly coupled with the polymers and thus more apparently affected everywhere in the flow. Recapitulating, at the interface the flow re-organizes itself. Vorticity close to the interface has to tilt and align parallel to it, the intermediate and the positive strain eigenvector becomes more strongly aligned along vorticity while the compressive one is more likely to be orthogonal to it compared to the bulk of the flow. The change in alignments are reflected in a distribution of enstrophy and strain production more skewed towards positive events of both, strain and enstrophy production. The shifted weight towards stretching (positive) events denotes how the interface is a region particularly active in

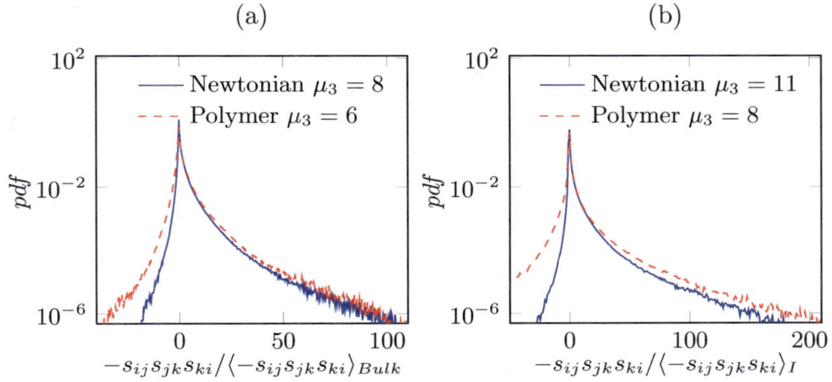

Figure 5.20: Probability density function of the normalized strain rate production $s_{ij}s_{jk}s_{ki}$ normalized by its average value. (a) in the bulk (b) at the interface, μ_3 is the estimated skewness of the distribution. Adapted from [16].

amplifying and redistributing fluctuations towards smaller scales. Polymers at the interface are found to re-orient preferably along the vorticity direction (i.e. parallel to the interface) and are also aligned with the local vortex stretching vector. The observable effect on strain and enstrophy production dynamics is to oppose the suppression of negative events observed for the Newtonian flow at the interface while leaving the occurrence of positive ones relative unaltered. The bulk, in comparison, sees changes only for the distribution of the negative strain rate production while the *pdf*s of enstrophy production are almost the same. Therefore, there is an apparent local effect of polymers at the interface over enstrophy mechanics that tends to oppose the effect of vortex stretching and surely contributes to the observed increase of the scales of the interface's features. For the case of the enstrophy production, the effect of polymers can be expected to be linked to their preferential alignment with the vortex stretching. Polymers are either opposing vortex stretching while extending or enhancing vortex compression while relaxing. This, though, is hardly assessable through Eulerian statistics only and this is where Lagrangian analysis can help.

5.2 Lagrangian Statistics

The Lagrangian statistics have been obtained by seeding the domains of the same 10 simulations with tracers and sampling the integrated position of the particles after fixed time intervals. Along the trajectories, also the local speed, velocity gradient tensor and polymer conformation tensor have been sampled. 1000 tracers have been seeded in each run for a total of $100'000$ sampled trajectories for a sampling time of about $30\tau_\eta$ after the steady state is reached. The focus of this analysis are the statistics of trajectories crossing the interface, hence tracers have been added only in regions within a distance between $1M$ and $3M$ from the centre of the domain and spanning its whole $y-z$ cross-section. This allows to have a good concentration of particles in the region around the turbulent/non-turbulent interface, increasing the number of interface crossings detected for a given number of seeded particles. Conditional

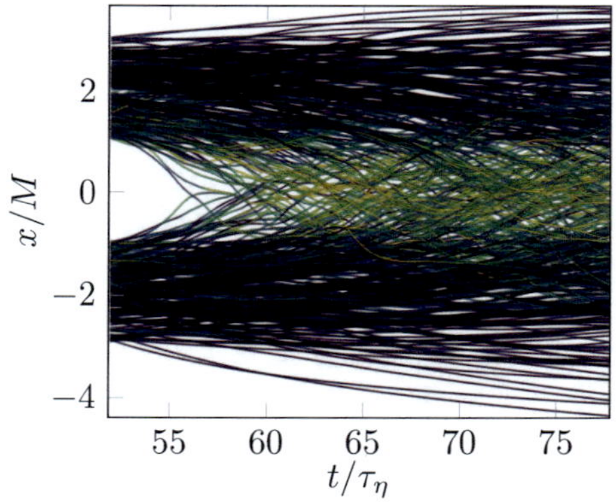

Figure 5.21: Position in time of a sub-sample of the Lagrangian tracers, the colour encodes the enstrophy in logarithmic scale.

Lagrangian statistics are performed by averaging together samples among all the trajectories with the same distance in time from the instant t_I at which the interface has been crossed. The crossing events are identified by means of enstrophy thresholding as done before for the interface detection in the Eulerian statistics. In this way two kind of crossing events can be univocally discerned: in the first case a trajectory that starts with zero vorticity is identified to be in the irrotational region and interface crossing t_I is identified at the first instant at which the enstrophy overcomes the selected threshold. These events are tagged as entrainment events. A second type of event is identified when the enstrophy of a trajectory is at some point in time above the threshold but it decays and remains below it for the remaining time of the simulation. In this case the last time at which the enstrophy is found above the average is identified as the interface crossing time t_I and the event is tagged as extrainment. This strategy permits to easily tag, separate and average together a subset of all the entrainment and extrainment events with the limitation of failing to detect all those events where the same trajectory crosses the interface in more time points. The dataset of sampled trajectories is pruned to accept only trajectories with a length of at least $\pm 10\tau_\eta$. Entrainment and extrainment events, especially in the polymer case, have different mechanics and the Lagrangian statistics permit to better discern their contribution to the Eulerian statistics observed in the previous section.

5.2.1 Entrainment

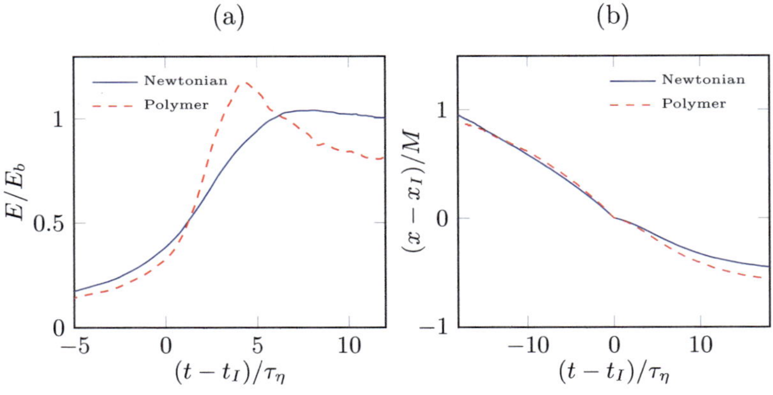

Figure 5.22: Conditional averages in time of entrainment events with respect to the distance from the interface crossing. Energy normalized by the energy in the bulk.

First the statistics of trajectories undergoing entrainment are considered. In Figure 5.22 (a) the conditional energy profiles with respect to the time t_I at which the interface is crossed are shown. In this and in the following plots $(t - t_I)/\tau_\eta = 0$ represents the point in time where the interface is crossed; negative times refer to times before the interface crossing, hence to the irrotational region. Positive times, on the other hand, refer to the turbulent region. It is possible to see in for $(t - t_I)/\tau_\eta < 0$ that the initial growth of the energy during the entrainment phase is relatively similar for both cases with slightly less energy for the polymer case. Nevertheless after the interface crossing energy rises more steeper in the polymer flow until it peaks at $(t - t_I)/\approx 5\tau_\eta$ and under the turbulence-reducing effect of the polymers it drops again until it reaches a local plateau. In order to have an impression of the distances travelled by the particle and the location in the flow of the observed features of the energy profiles, it is interesting to consider the average position of the particles from the interface. Figure 5.22 (b) depicts the average position of the particles respect to the position x_I at which its trajectory crosses the interface. Here positive distances represent the irrotational region, while negative ones represent the turbulent one. For $(t - t_I)/\approx -20\tau_\eta$ particles start on average at about $1M$ from the position of interface crossing and approach the interface with a quasi-constant velocity in x-direction. After the interface crossing, despite the observed increase of kinetic energy, the particles seem to proceed at a slower pace towards the bulk of the flow, confirming the observations of previous studies [98]. This is valid for both, the Newtonian and the polymer case, even though on average polymers appear to have a slightly higher speed towards the bulk. Figure 5.23 (a) shows how the enstrophy increases and reaches a plateau

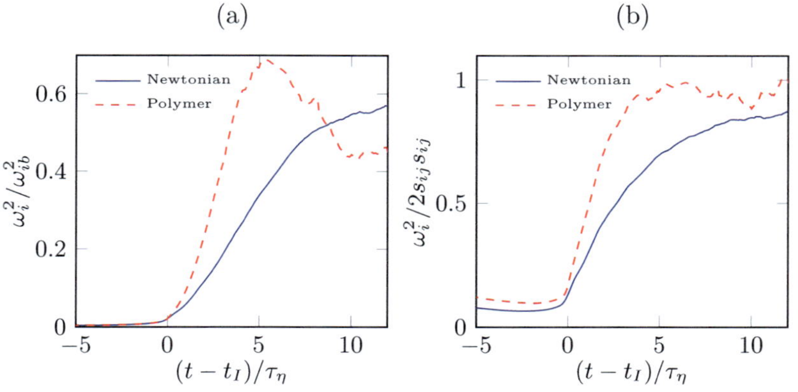

Figure 5.23: Conditional averages in time of entrainment events with respect to the distance from the interface crossing. (a) Enstrophy normalized by the enstrophy in the bulk, (b) ratio between enstrophy and strain.

in both cases, similarly to what has been observed before in the Eulerian conditional statistics. The plateau is reached in both cases within $10\tau_\eta$ from the interface crossing, but for the polymer case it is possible to observe a sharper gradient and a distinct peak of enstrophy at about $5\tau_\eta$. The peak is the mark of the transition of the polymers to a stretched state. At this point they start to react back to the flow and suppress enstrophy. Again, similarly to the conditional Eulerian statistics, the ratio between enstrophy and strain in Figure 5.23 (b) shows a steeper slope for the polymers compared to the Newtonian case, confirming that for polymers vorticity across the interface reaches the equilibrium with strain faster than in the Newtonian flow. The transition of

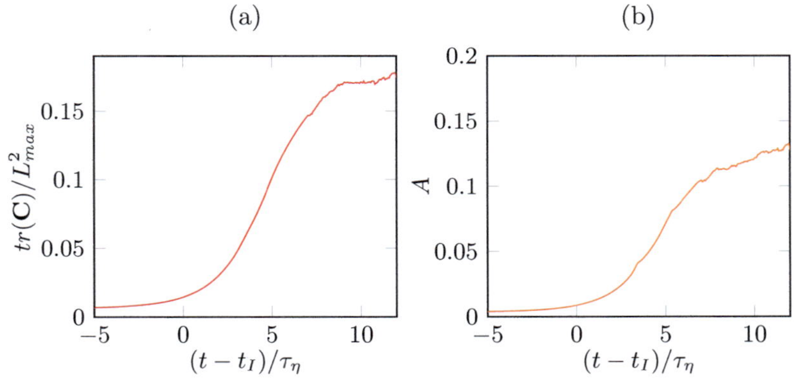

Figure 5.24: Conditional averages in time of entrainment events with respect to the distance from the interface crossing. (a) polymer extension normalized by average extension in the bulk, (b) free energy A_p.

polymers to the stretched state is easily observable in the conditional profile of polymer stretching in Figure 5.24 (a) which reach a plateau around the same time at which enstrophy does. A plateau is observed at around $10\tau_\eta$, where polymers reach and average extension of around 17% of the maximum allowed one. Furthemore it can be noted how the polymer in the irrotational region have almost negligible extension and they start to significantly stretch only within the turbulent region. Figure 5.24 (b) depicts the energy stored in the polymers as free energy A_p. The energy stored in the polymers keeps growing. The plateaus observed in the Lagrangian entrainment statistics are consistent with observations in previous studies that stated that entrained particles tend to station in the whereabouts of the interface for some time [96, 98].

5.2.2 Extrainment

It is interesting to observe how the extrainment process, which also affects the mechanics of the turbulent/non-turbulent interface, unfolds. In the following set of plots negative $(t - t_I)/\tau_\eta$ are portions of trajectories that are within the turbulent region while positive $(t - t_I)/\tau_\eta$ represent portions of trajectories in the irrotational region. The turbulent kinetic energy shown in Figure 5.25 (a)

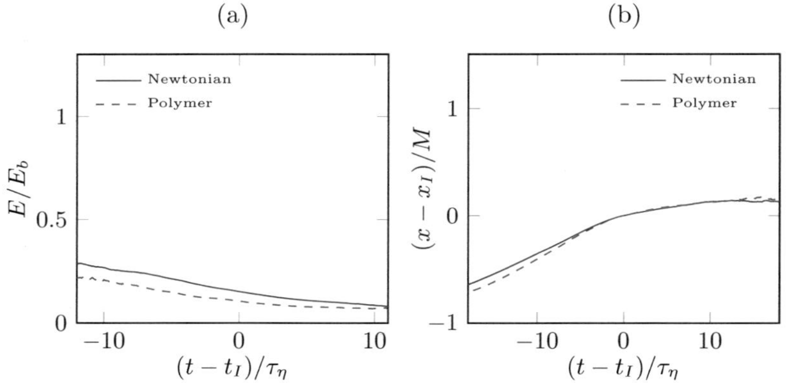

Figure 5.25: Conditional averages in time of extrainment events with respect to the distance from the interface crossing. Energy normalized by the energy in the bulk.

appears to be maintained during the extrainment relatively flat gradients for both the polymer and the Newtonian case, the main difference being a lower level of kinetic energy for the polymer one. Similarly to the entrainment case there is no direct correlation between turbulent kinetic energy and velocity respect to the interface position. Indeed, by looking at Figure 5.25 (b) one can see how tracers in the polymer case starts slightly further away from the interface respect to Newtonian tracers indicating slightly larger speeds towards the interface. Once crossed the interface the tracers appear to station just outside it reaching within $10\tau_\eta$ a negligible speed in x-direction. From the distance statistics of both entrainment and extrainment events it appears that particles after crossing the interface tend to remain in its whereabouts and especially there is a delay between the entrainment of a particles and its actual mixing into the bulk of the flow. The introduction of polymers seems to slightly increase the exchange of particles between bulk and interface by increasing both the speed of trajectories towards the bulk during entrainment as well as the speed of trajectories towards the interface during extrainment. As can be observed from Figure 5.26 (a) on average the sampled trajectories of

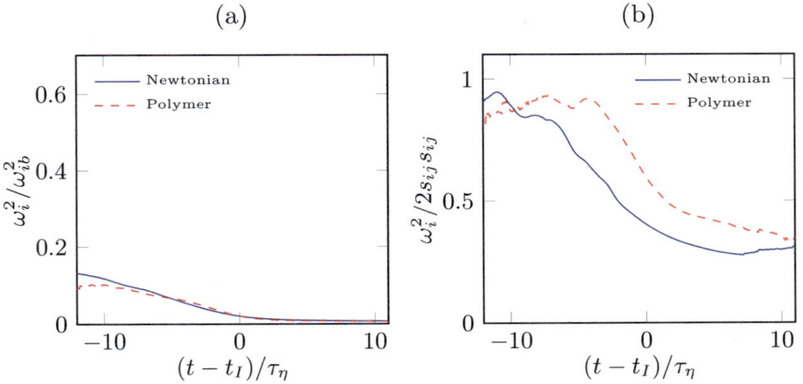

Figure 5.26: Conditional averages in time of extrainment events with respect to the distance from the interface crossing.(a) Enstrophy normalized by the enstrophy in the bulk, (b) ratio between enstrophy and strain.

the particles undergoing extrainment show similar profiles of enstrophy for the Newtonian and the polymer case. Interestingly the transition from the turbulent to the non-turbulent region is demarcated by only a light change in the slope of the profiles. While enstrophy profiles appear almost unchanged, polymers still affect the strain. Indeed, due to a faster decay of strain fluctuations the ratio $\omega_i^2/2s_{ij}s_{ij}$ of Figure 5.27 (b) is larger compared to the Newtonian case. Looking at the average polymer extension in Figure 5.27 (a) it can be noticed how this is lower than the threshold of activation observed for the entrainment statistics of Figure 5.23, so it can be expected that polymers are not the major responsible for the decay in enstrophy of these particles. Moreover, a small residual extension is maintained for more than $10\tau_\eta$ after the crossing of the interface confirming that what observed for the non-turbulent region in the Eulerian conditional statistics of Figures 5.11 is partly imputable to the expulsion of stretched

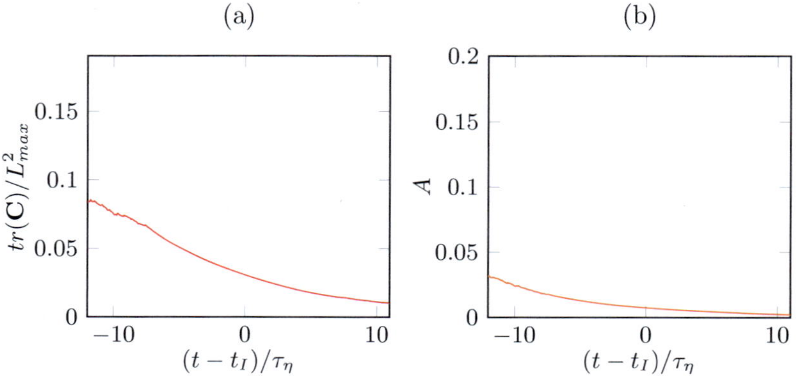

Figure 5.27: Conditional averages in time of extrainment events with respect to the distance from the interface crossing.

particles from the turbulent region. The free energy in Figure 5.27 (b) follows the decay of the average extension and similarly retains some non-zero level quite far from the interface crossing. The Lagrangian analysis is concluded

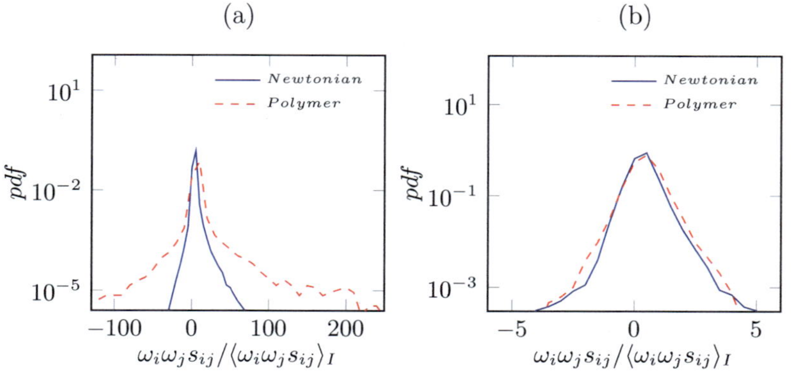

Figure 5.28: Probability density function of the normalized enstrophy production from the Lagrangian tracers at the interface crossing. (a) Entrainment events, (b) extrainment.

with a comparison of enstrophy production dynamics between entrainment and extrainment events in the proximity of the interface. Indeed, the plots of Figure 5.28 depict the probability distribution of the enstrophy production normalized by its average value at the interface of the Lagrangian tracers for which $t-t_i = \pm\tau_\eta$. In the case of entrainment events, the normalized production shows a much larger probability of extreme events for the polymer case. In both cases the distribution maintains a prevalence of positive events over the negative ones and, overall, they both positively contribute to the enstrophy variation at the interface. For the case of extrainment events, as expected, much weaker enstrophy production is observed, the distributions are flatter and more similar to Gaussians than the ones previously observed even though a little prevalence of positive enstrophy production is still present. More noticeably, both the distributions are very similar further showing that both Newtonian and Polymer trajectories undergo to similar decay histories. When observing these *pdf*s, it is hard to draw a parallel with their Eulerian equivalent. In particular, from these distributions, it is not possible to detect the shift towards more negative events observed for the Eulerian statistics with polymers. Similarly, caution should be used in trying to link the Lagrangian conditional profiles of the statistics shown here to their Eulerian counterparts. It is obvious that the entrainment and extrainment events brings very different contributions to the global statistics of the interface. The extrainment statistics depict another side of the interface, where properties of the two regions blend smoothly and the border between the two regions is more blurred. The sharp gradients across the turbulent/non-turbulent interface observed in the Eulerian conditional statistics in this sense appear to be more related the Entrainment events. Nevertheless, at the steady state it can be expected that entrainment events are as likely and equally important as the extrainment ones. Unfortunately, a balance between the two could not be performed being the sampling algorithm capable to track only the first (entrainment) and the last (extrainment) interface crossing for any given trajectory. The majority of interface crossings are from particles that after entrainment/extrainment remain or return to the whereabouts of the interface and undergo to extrainment/entrainment many times along the same trajectory. As evidence suggests, both entrained and extrained particles tend to remain close to the interface, and these events can be expected to consist of a relevant number of all the entrainment/extrainment events at the steady state. Indeed, all the events tagged either as entrainment or extrainment by the algorithm used here represents around 21% of all the detected crossings of the enstrophy threshold for the Newtonian case and 17% for the polymer case. In total around 79 – 83% of the events remain untagged and represent all those cases where the threshold is crossed due to either internal turbulent fluctuations of enstrophy or actual multiple crossing of the interface by the same trajectory. The Lagrangian statistics are hence representative of only a limited subset of the trajectories crossing the interface and cannot explain the Eulerian statistics shown in the previous sections alone. Nevertheless, it is interesting to study the differences between the entrainment and extrainment processes as these may be more or less predominant during different phases of the life of a turbulent patch. For example, during the growing phase at the interface, the entrainment events will prevail leading to different global properties compared to the steady state case. Conversely, in a receding flow the extrainment events will be determinant to the interface properties. These factors are particularly important in order to understand the behaviour of the polymer model in the transient phases of the turbulent patch. Extrainment showed to be the strongest contribution to the observed presence of stretched polymers in the non-turbulent region. At the steady-state these can accumulate producing a shell around the irrotational side of the interface of uncoiled polymers capable to apparently affect the strain field in that region. Being currently unable to address the transient phase of the polymers, it is not possible to determine if the observed mechanics of polymer re-organization of strain/vorticity alignments and their effect on the strain/enstrophy production can be generalized also to the growing phase of the turbulent patch.

6 Conclusions

In this thesis the relatively unexplored topic of the mechanics of turbulent propagation in dilute polymer solutions has been studied with the objective of increasing the current understanding on turbulent propagation not only of this particular kind of flow but also of Newtonian turbulent flows in general. Among the many open issues in the understanding of turbulent propagation there is the one of properly assessing the two dichotomies of small versus large scales role and of viscous versus inertial dynamics. Polymers in this framework permit to study the propagation of turbulence in conditions where inertial dynamics and scale distribution have been altered compared to Newtonian turbulence, without affecting the physical mechanism of molecular viscous diffusion. Preliminary spectral analysis on shearless turbulent/non-turbulent interfaces of a Newtonian decaying flow highlighted the presence of a complex system of viscous and non-viscous fluxes, with sources and sinks in different regions of the flow and at different turbulent scales. In particular a central role of large-scale inhomogeneities in the flow has been found. The inhomogeneities prompt advective fluxes of enstrophy from the bulk towards the interface which feed turbulent fluctuations near the interface and further the propagation and enstrophy growth despite the general decay in the rest of the flow. It has been observed that while conditional statistics in literature confirm the presence of a thin layer dominated by viscous diffusion [79], the spectral analysis found that scales parallel to the interface of such regions are relatively large. Qualitative observations of the viscous diffusion of enstrophy near the interface confirm the presence of waves of alternating positive and negative diffusion with thin thickness normally to the interface but with relatively large extension parallel to it.

DNS of turbulent fronts in a continuously forced set-up have been performed with Newtonian and FENE-P models for the constitutive relation between stress and deformation. The employing of a coarse grained model as the FENE-P is imposed by the current limitations on the computational power available and requires a validation against experimental observation in order to confirm the trends observed in the simulations. A number of difficulties have been encountered during this study that might limit the generality of the observations made. First, both experiments and simulations with polymers present some sensitivity to the parameters chosen and results cannot be generalized to all the parameter space. The limits of both, the experimental and numerical technique limited the flow to small Reynolds numbers and again polymers are known to lead to contradictory behaviours with increasing Reynolds numbers [51] partially due the fact that polymers interact differently with the different turbulent structures that can arise at different Reynolds numbers. On another side Reynolds number's effects are expected to be less influential on a region like the turbulent/non-turbulent interface where by definition turbulent fluctuations are almost completely decayed.

Both simulations and experiments confirmed a reduction of the maximum propagation of the turbulent region for the dilute polymer case, even with comparable levels of turbulent fluctuations respect to the Newtonian case. Also, both experiments and simulations showed a reduction in the fractal-like complexity of the interface for the polymer case, leading to flatter interface featuring larger scales. Unfortunately it has not been possible to study the growth rate of the turbulent region for the polymer case due to the fact that the FENE-P model requires few eddy turnover times in order to reach sufficient average extensions to start to affect the flow. Despite its deficiencies in properly predicting the transient behaviour, the FENE-P model proved itself again to qualitatively predict the behaviour of dilute polymer solutions even in this non-canonical case. The dynamics of vorticity and strain at the turbulent/non-turbulent interface have been extensively investigated permitting to observe how those are altered by the introduction of the polymeric stress. Like many other anisotropic Lagrangian tracers, polymers have been found to assume the orientation of the local vorticity, leading to a direct interaction with the vortex stretching. In those regions of the flow that are still approximatively isotropic this does not lead to appreciable differences in the distribution of positive and negative events of the normalized enstrophy production $\omega_i \omega_j s_{ij} / \langle \omega_i \omega_j s_{ij} \rangle$ as polymers seem to affect both kind of events in the same way and a üniversal"balance is conserved. At the interface though, the balance moves for dilute polymer solutions towards more intense compressive events. This globally leads towards lower production of enstrophy at the interface, shift towards larger scales, reduced convolution of the interface and entrainment flux.

Through Lagrangian tracking it has been possible to separate trajectories in three different groups: trajectories that enter the turbulent region and remain within it, trajectories that leave the turbulent region without returning turbulent again and, finally, trajectories that cross the interface several times in both directions. It has been possible to univocally identify and tag only the first two type of trajectories, but those already showed how particles crossing the interface from the irrotational region towards the turbulent one have dramatically different statistics respect to the ones crossing the interface in the opposite direction. In particular entrainment appears to be responsible for the relatively steep gradients observed in conditional Eulerian statistics of many quantities. On the opposite side extrainment events have very flat profiles that tend to smooth the aforementioned gradients. It can be argued that the balance between entrainment and extrainment events defines the Eulerian statistics of the interface partially explaining some of the differences observed between different type of flows or even between different regions of the same flow [8, 95]. The distinction between the extrainment and entrainment Lagrangian statistics is even more marked for the polymer case. Indeed, since polymers are in the coiled conformation at the beginning of the entrainment, they only little affect the enstrophy at the interface, allowing for faster growths in the more energetically stirred polymer flow. On the other hand, extrainment events have very similar enstrophy and energy profiles and areresponsible for the observed presence of residual polymer stretching in the non-turbulent region. The residual stretching of the polymers after extrainment might be significant for the largest class of events represented by trajectories crossing the interface multiple times. Along those trajectories, polymers can constantly maintain a certain degree of extension and react more rapidly during subsequent entrainment events. In wall-bounded flows, polymers have been observed to strongly interact with near-wall turbulent structures. Near the wall polymers are indeed more intensely affecting the flow [34], they are strongly aligned with coherent vortical structures and increase the anisotropy of turbulent fluctuations, while far from the wall they have a more isotropic behaviour [29, 51]. Similarly, for the flow investigated in this thesis the polymers have more freedom of orientation in the bulk of the flow where the properties of both polymers and flow better approximate the ones of homogeneous isotropic turbulence.

Concluding, it has been possible to obtain some further insight on the mechanism with which polymer affect turbulence and interact with coherent structures in the flow. At the border between the turbulent and the irrotational region, it is possible to see a strong alignment with the large (though thin) organized structure which is the turbulent/non-turbulent interface. A preferential alignment of polymers with large coherent structures means that the polymer stresses can affect larger turbulent scales at the interface compared to the ones affected in more turbulent, chaotic and isotropic regions of the flow. These observations can probably be extended to flow control applications, where the control might be more easily applicable and more effective in those regions of a flow where turbulence experiences a reduced degree of freedom and is forced to organize in large structures.

A Appendix

A.1 Numerical Implementation

The incompressible Navier-Stokes equations and the evolution of the polymer conformation tensor are discretized by means of a pseudo-spectral method based on Fourier series and are integrated in time through a partially implicit Crank-Nicholson/Runge-Kutta scheme following the implementation used in De Angelis et al. 2005 [21].

The momentum conservation is given by:

$$\frac{\partial u_i}{\partial t} + u_j \frac{\partial u_i}{\partial x_j} = -\frac{\partial p}{\partial x_i} + \frac{1}{Re} \nabla^2 u_i + g_i + f_i, \tag{A.1}$$

where f_i is the ith-component of the body force and g_i is the ith-component of the added polymer contribution

$$g_i = \frac{1}{Re} \frac{\partial T_{ij}^p}{\partial x_j}. \tag{A.2}$$

Using the FENE-P model the stress $T_i j$ is given by:

$$T_{ij}^p = \frac{\eta_p}{\tau} \left(\frac{L_{max}^2 - 3}{L_{max}^2 - Tr(C_{ij})} C_{ij} - \delta_{ij} \right). \tag{A.3}$$

Here τ is the relaxation time of the polymer chain, η_p is the ratio between the asymptotic zero-shear-rate viscosity of the solution with polymers and the solvent viscosity. L_{max}^2 is the maximum allowed extension of the polymer chains and C_{ij} is the conformation tensor that represents the average orientation and extension of the polymer chains at a given point of the field. The evolution of the conformation tensor in the current implementation is governed by

$$\frac{\partial C_{ij}}{\partial t} + u_k \frac{\partial C_{ij}}{\partial x_k} = -\frac{1}{\tau}\left(\frac{L_{max}^2 - 3}{L_{max}^2 - Tr(\mathbf{C})} C_{ij} - \delta_{ij}\right) + \frac{\partial u_i}{\partial x_r} C_{rj} + C_{ir}\frac{\partial u_j}{\partial x_r} + \chi \nabla^2 C_{ij}, \quad \text{(A.4)}$$

where the diffusive term $\chi \nabla^2 C_{ij}$ is added in order to increase the stability of the simulation. The Navier-Stokes equations when transformed in the Fourier space assume the form

$$\frac{\partial \hat{u}_i}{\partial t} = \hat{h}_i - \frac{k_i}{k^2}(k_i \hat{h}_i) + \frac{1}{Re}\nabla^2 \hat{u}_i + \hat{g}_i - \frac{k_i}{k^2}(k_i \hat{g}_i) + \hat{f}_i - \frac{k_i}{k^2}(k_i \hat{f}_i) \quad \text{(A.5)}$$

where $\hat{h}_i = u_j \hat{\omega_k} \epsilon_{ijk}$. The Equation A.5 is then integrated in time using a third order Runge-Kutta scheme in the form:

$$\hat{u}_i^{n+1} = \hat{u}_i^n + a_n \hat{H}_i^n + b_n \hat{H}_i^{n-1} - \frac{a_n + b_n}{2Re} k^2 \left(\hat{u}_i^{n+1} + \hat{u}_i^n\right) + \frac{a_n + b_n}{2Re}\left(\hat{B}_i^{n+1} + \hat{B}_i^n\right) \quad \text{(A.6)}$$

Where $\hat{B}_i = \hat{g}_i - \frac{k_i}{k^2}(k_i \hat{g}_i) + \hat{f}_i - \frac{k_i}{k^2}(k_i \hat{f}_i)$ and $hatH_i = \hat{h}_i - \frac{k_i}{k^2}(k_i \hat{h}_i)$. The non-linear term is integrated using an Adam-Bashforth scheme and the linear term by using an implicit Carnk-Nicolson one.

Lagrangian statistics have been computed by seeding the flow at random coordinates within a desired region of it. The particles' positions are then integrated at each time step using the Eulerian velocities. Being in general the particles' positions not coincident to the discretized points of the domain the particle velocity is obtained by a bicubic interpolation of the velocities of the neighbouring points. The same applies for all the other sampled quantities at the particles' positions.

Table A.1: Parameters for the characterization of the forcing. All simulations were run at $Re = 200$, $\Delta t = 0.001$.

Name	K	Δ	T_f	M	u'	Re	λ	Re_λ
Base	3.8	0.065	0.1	$L_{y-z}/8$	0.50	80	0.65	65
A1	5.8	0.065	0.1	$L_{y-z}/8$	0.58	91	0.61	70
A2	6.8	0.065	0.1	$L_{y-z}/8$	0.65	102	0.59	76
A3	3.8	0.065	0.1	$L_{y-z}/8$	0.75	117	0.56	84
T1	3.8	0.065	0.05	$L_{y-z}/8$	0.71	112	0.59	84
T2	3.8	0.065	0.3	$L_{y-z}/8$	0.38	59	0.70	53
D1	3.8	0.075	0.1	$L_{y-z}/8$	0.52	82	0.65	68
M1	3.8	0.065	0.1	$L_{y-z}/4$	0.50	157	0.76	76
M2	3.8	0.065	0.1	$L_{y-z}/16$	0.44	34	0.46	41

A.2 Forcing characterization

The thickness Δ is usually chosen to be $\Delta \approx M$ in order to have a more isotropic energy injection scales. With our forcing model the governing parameters are the maximum amplitude A_0, the forcing interval Δt_u, the mesh-size M and Δ. A number of simulations have been performed with the purpose of assessing the effect of each of these parameters and it has been observed how, within the range used, the forcing produced results consistent with the ones expected from oscillating grid turbulence. The data-set presented here refers to a single simulation re-initialized every time with different forcing parameters. All simulations, apart the ones at different forcing frequency $1/T_f$ and different mesh size M, share the same random number time-sequence for the generation of the forcing distribution. This has been made in order to allow a direct comparison between single runs. The forcing parameters used in this parametric study are recapped in Table A.1 where also some flow statistics, sampled in the middle of the forced region, are presented. Figures A.1–A.2–A.3–A.4 show the effect of the choice of the various parameters over the average energy in the middle of the forced region and over the interface position detected by a threshold of enstrophy equal to the 2% of the bulk enstrophy. While the growth phase of the patch seems to be accelerated by parameters that increase the energy content of the forced region the final

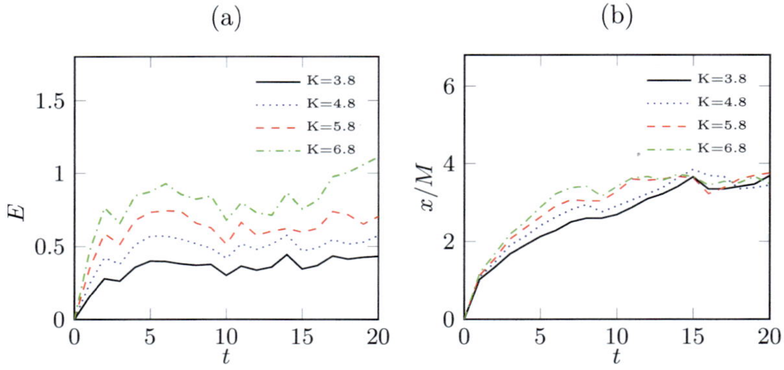

Figure A.1: Kinetic energy in the forced region(a)and average interface position (b).

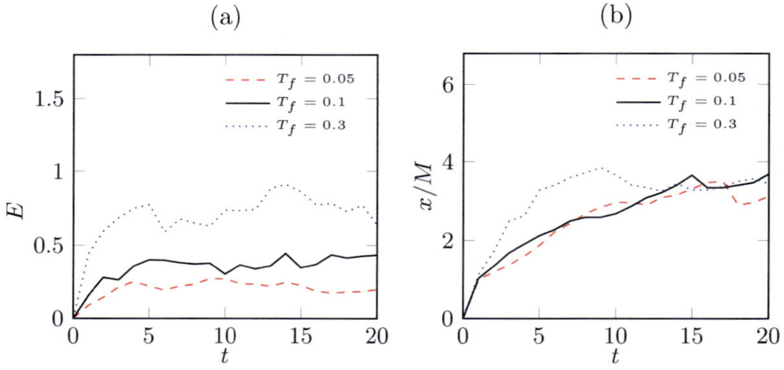

Figure A.2: Kinetic energy in the forced region(a)and average interface position (b).

patch size appears to be less sensitive to changes in the forcing parameters. According to this metric the parameters that affect the most the final position of the turbulent front are the energy injection scales M and Δ.

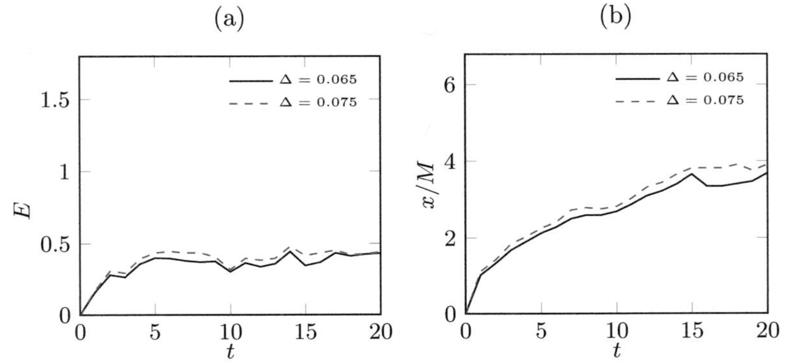

Figure A.3: Kinetic energy at the center of the forced region (a), average interface position (b).

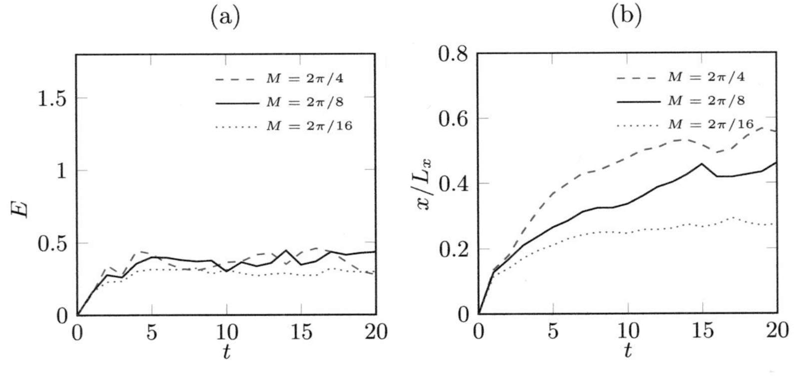

Figure A.4: Kinetic energy at the center of the forced region (a), average interface position (b). In this case for the sake of comparison the interface position has been normalized by the box half size L_x

Table A.2: Parameters for the test of the polymer parameters. All simulations were run at $Re = 200$, $\Delta t = 0.001$.

Name	L^2_{max}	τ	η_p	χ	u'	Re	λ	Re_λ
Base	5000	2	0.1	1/200	0.45	70	0.77	69
ET1	5000	2	0.06	1/200	0.47	73	0.74	70
ET2	5000	2	0.25	1/200	0.44	69	0.79	68
TA1	5000	0.8	0.1	1/200	0.43	68	0.75	64
TA2	5000	4	0.1	1/200	0.41	66	0.72	60
TA3	5000	7	0.1	1/200	0.50	78	0.72	71
P1	5000	4	0.25	1/200	0.41	64	0.72	62
P2	3600	5	0.25	1/200	0.36	62	0.76	60
CHI	5000	2	0.1	1/100	0.45	70	0.47	69

A.3 Choice of the parameters for the FENE-P model

Within a limited range of values, also the effect of the choice of the FENE-P model parameters has been assessed. A set of simulations have been performed with fixed forcing parameters and different parameters for the polymer model. The simulations have all the same random number time-sequence for the generation of the forcing distribution in order to have a better comparison. The FENE-P parameters used in this parametric study are recapped in Table A.2 where, like previously done for the forcing study, also some flow statistics sampled in the middle of the forced region are presented. The simulations have been stopped at the time at which turbulence for the base case stops propagating. In Figures A.5–A.10–A.11–A.12 show like for the Newtonian study little sensitivity of the detect interface position with respect to the model parameters. For the cases TA1,TA2,TA3 in Figure A.10, a consistent trend of reduction of the kinetic energy could be found at increasing values of τ. The same leads to contradictory results for the interface propagation, where all three cases led to larger patches than the base case. Simulations P1 and P2 test sets of parameters where the value of K_{coiled} is kept constant where:

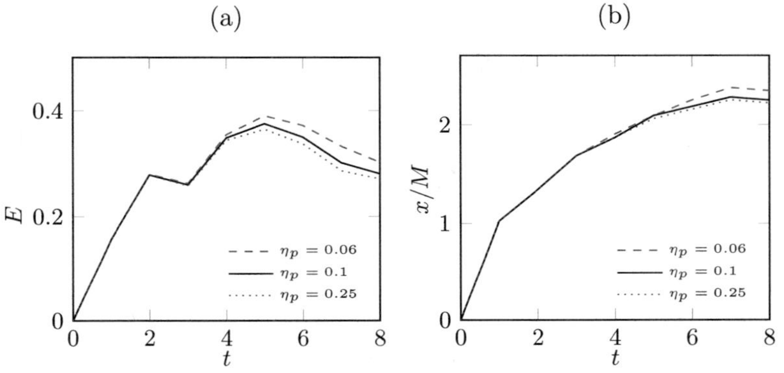

Figure A.5: Kinetic energy at the center of the forced region (a), average interface position (b).

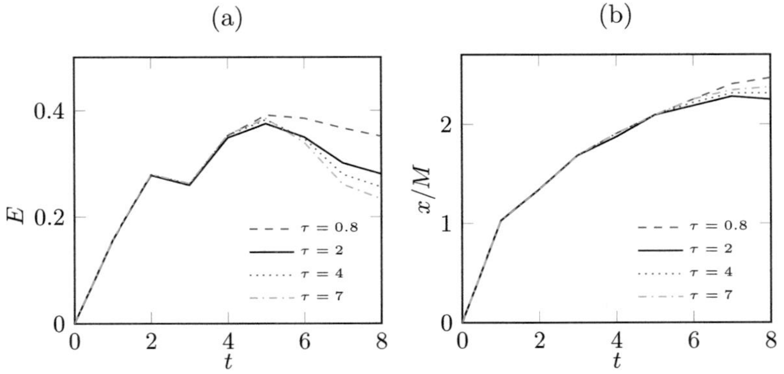

Figure A.6: Kinetic energy at the center of the forced region (a), average interface position (b).

$$K_{coiled} = \frac{\eta_P}{\tau} \frac{L_{max}^2 - 3}{L_{max}^2 - Tr(C_{ij})} \tag{A.7}$$

is computed for small extensions (below 1% of the maximum allowed). Therefore, these simulations have almost the same polymer reaction force in the small extension range. These simulations, though having the largest reductions

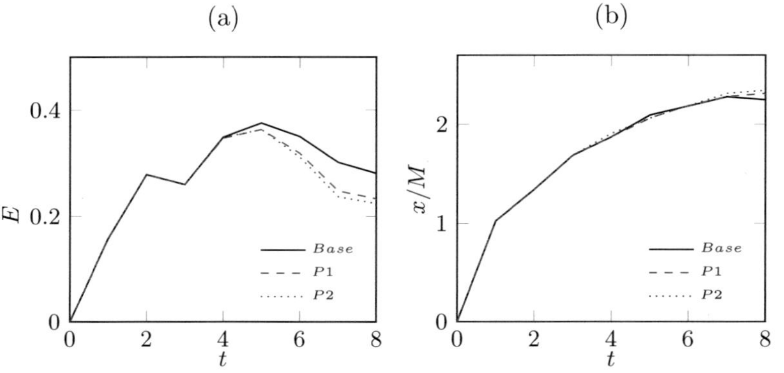

Figure A.7: Kinetic energy at the center of the forced region (a), average interface position (b).

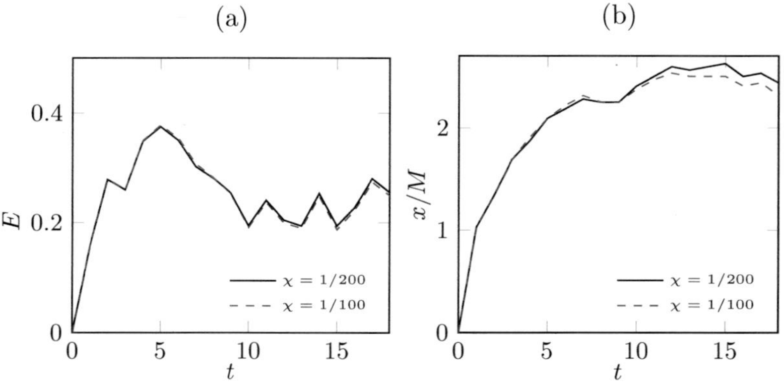

Figure A.8: Kinetic energy at the center of the forced region (a), average interface position (b).

in the kinetic energy in the tested parameter space, lead to very similar profiles of the patch growth further underlying the importance of the low extensions regime for the FENE-P model in turbulent/non-turbulent interfaces. Finally, simulation CHI has been realized in order to test the effect of the artificial diffusivity on the energy and the propagation. This simulations show maximum differences below 2% for both, energy and propagation when the artificial diffusivity is double the one of the base case.

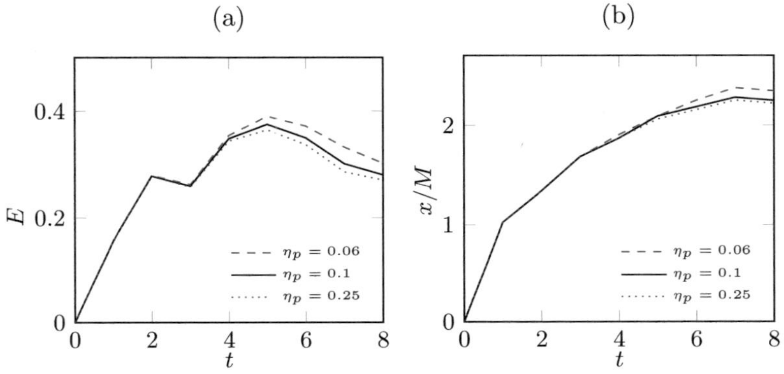

Figure A.9: Kinetic energy at the center of the forced region (a), average interface position (b).

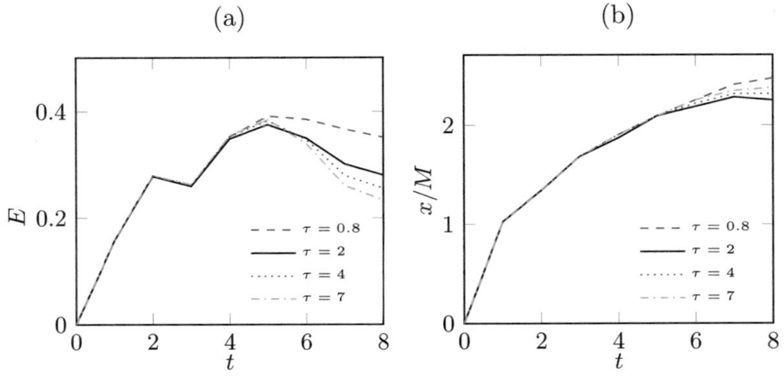

Figure A.10: Kinetic energy at the center of the forced region (a), average interface position (b).

A.4 Interface Detection Methods and conditional averages

Several methods can be found in literature that permit to track the position of the turbulent/non-turbulent interface. Each of them can lead to different detected interfaces, often producing results that are difficult to compare [8]. Some further detail on the interface detection techniques is given here in

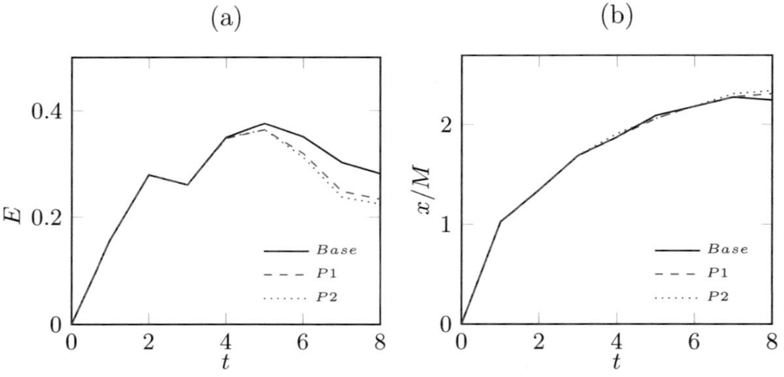

Figure A.11: Kinetic energy at the center of the forced region (a), average interface position (b).

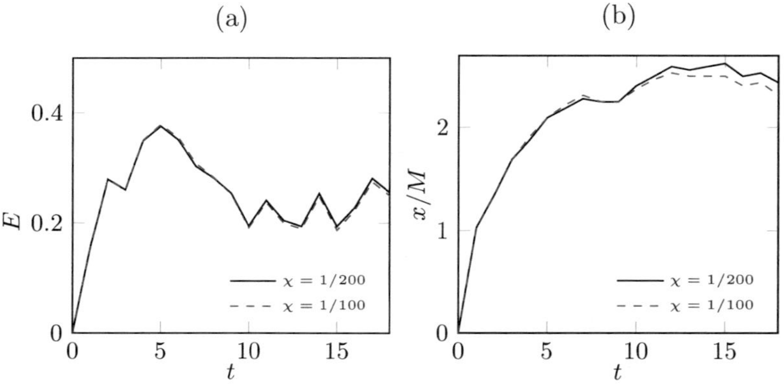

Figure A.12: Kinetic energy at the center of the forced region (a), average interface position (b).

order to better clarify the origin of some variability in the literature results. As it has been introduced in the first chapter of the thesis, several quantities present relatively sharp changes across the interface and, virtually, any of these quantities can be used for interface detection. Scalar quantities are particularly suitable for such purpose and, for example, velocity magnitude, enstrophy, temperature, concentrations of scalars (both passive and reacting) had been used for interface detection purposes [19, 36, 42]. The basic principle involves setting a maximum value or threshold of such scalar quantity, above which the fluid can be considered with a certain confidence to be turbulent. Then all the points of the field are tagged as turbulent when the scalar quantity is above the threshold, and non-turbulent when it is below. Further post-processing is required in order to extract the boundary between the two regions and a number of algorithms have been developed in order to do so. This has introduced a further source of variability in the interfaces detected by different studies. Part of the post-processing usually implies removing internal pockets of irrotational fluid trapped inside the turbulent region as well as isolated turbulent bubbles in the irrotational region detached by the mass of the turbulent flow. When this pocket/bubble remotion is applied to 2D slices of the scalar field, parts that in the more complex 3D field would result connected to the main irrotational/turbulent body are also removed. The error associated to this remotion is usually considered negligible and to the knowledge of the author only Borrell and Jimenez [8] implemented an algorithm that accounts for connected 3-dimensional regions at the cost of a sensible increase of the computational costs. In these study the turbulent/non-turbulent interface has been identified by mean of an enstrophy threshold, i.e, non-turbulent regions are identified in the flow where the vorticity magnitude falls below a certain level. Such approach has proved itself one of the most robust techniques for interface detection, though it presents the problem of the choice of a meaningful value for the threshold Ω_{th}. Following this method, the average position of the interface has been define as $X_I = \langle x_I \rangle$ where $x_I = x_I(y, z)$ is the instantaneous position of the interface detected by finding the outermost point where enstrophy, $\Omega = \omega_i \omega_i / 2$ with ω_i denoting vorticity, equals a given threshold. Figure A.13 shows a comparison between the interface detected by the algorithm used in the present work and the one detected by a simplified implementation of the algorithm used in Borrell and Jimenez 2016 [8]. As in both, experiments and simulations a certain level of noise is always present in the non-turbulent region, 0 level thresholds are not practically applicable. Moreover, it can be argued that not any arbitrarily low enstrophy fluctuation denotes turbulence. Hence, the choice for the threshold depends on empirical observations and on constraints dependent on the particular flow case. For example in certain time and space evolving flows a single constant value for the threshold would fail to properly detect the interface at every stage of the flow. In these cases the threshold should be dependent on some local-in-time or -space turbulent property. In the present work a time-dependent threshold has been used and, following Wolf et al. 2013 [98], a value of 2% of the mean enstrophy in the bulk of the flow at a given time has been used ($\Omega_{th} = 0.02\Omega_0$). Such a choice

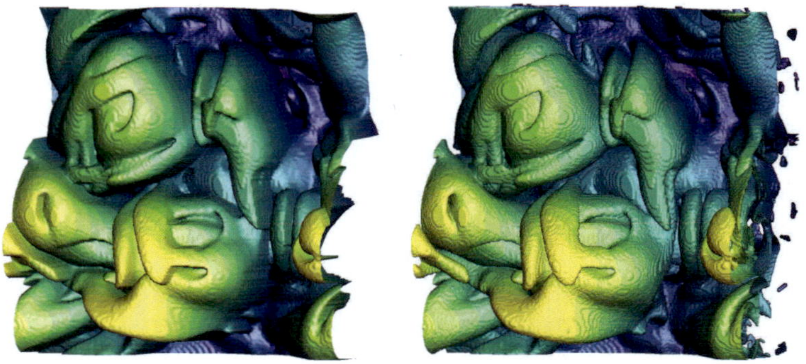

Figure A.13: Example of the interface by two different detection algorithms. Right: only the outermost thresholded points are selected. Left: simplified algorithm from Borrell and Jimenz algorithm prior the removal of pockets and bubbles.

has proven well suited through the study for identifying the region of the flow where propagation takes place. In Chapter I it permits to properly detect the interface in the location of maximum positive enstrophy variation of the single-points budgets. In chapter IV it properly detect the interface at the end of the region of linear growth of the integral scale and the region of maximum velocity anisotropy.

Once identified the interface, it is possible to define properties related to the distance from the interface. This consists in defining a new local reference system centered at the local position of the interface. Then, statistics are sampled averaging together all the points at the same coordinate respect to its local interface position. Different choices of this reference system can give different results, especially at growing distances from the interface. The simplest choice is to use the horizontal distance Δx referred to the outermost detected interface position for every coordinate couple (y, z). In doing so, the local reference system maintains the same angle respect to the global one, and changes its position only. This has the disadvantage of hiding the contribution from inner parts of the interface, but it permit to depict statistics at the interface as well as to return to the fully turbulent (classical) statistics far enough from it. This is the approach used here and in Chauhan et al., Westerweel et al. and da Silva and Taveira [11, 78, 97]. A second approach consists in defining a local reference system oriented as the direction normal to the local interface. This approach is used for example in Watanabe et al. 2016 [96] and da Silva and Pereira [73].

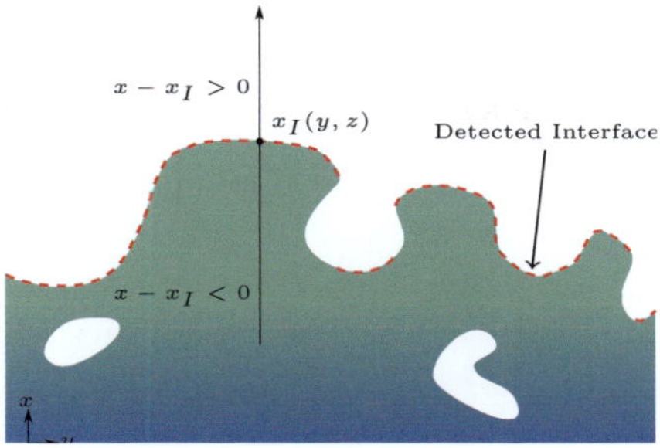

Figure A.14: Conditional averaging procedure.

This technique improves the quality of the statistics near the interface, especially for strongly convoluted ones, but the local axes tend to intercept each other further away from it, making the samples at larger distances far from homogeneous. Borrell and Jimenez 2016 [8] use a third approach in which, for every point p of the field, they find the ball distance Δ_b from the interface defined as the minimum radius of a sphere with center in p and surface tangent to the interface. Then they compute statistics as a function of the distance Δ_b. The operation of generating the new reference systems for all the points of the field has computational cost of $N_p log(N)$ and was deemed to expensive for the application of this study. As already pointed out, the three approaches give in general different results mostly in the mid to far distance to the interface. The more regular is the interface, though, the more similar are the statistics obtained by the three methods. Another source of arbitrariness in the compute statistics is the choice of the enstrophy threshold. In experiments, a value is chosen such that it is preferably of some orders of magnitude lower than average values in the bulk but still robustly above the magnitude of the noise in the non-turbulent region [42]. In such a way, there is a range of thresholds for which the turbulent volume enclosed by the detected interface do not changes much [19]. Borrell and Jimenez, though, pointed out that this plateau spans over several decades of Ω and that both, turbulent statistics and the topology of the interface can change considerably within this range. They suggest, nevertheless, an order of magnitude for the choice of the interface [8]. Here, the threshold is time dependent and defined as the a fraction of the average enstrophy of the bulk. The position at which this average is computed corresponds to a plane at a distance of $x = 0.6M$ from the middle of the domain, which is just outside the forced region. The chosen threshold equals 2% of the average bulk enstrophy, such a value roughly correspond to the threshold used for similar flows and Re_λ in [42, 43, 98], and an half of the one used for jet flows in [7, 95]. The interface is highly fluctuating and in both, simulations and experiments, can reach deep into the turbulent bulk [8, 19]. In order to avoid effects of the body force on the statistics, all those samples within the forced region have been eliminated from the computation. For the Lagrangian statistics the conditional averaging is performed with respect to the time. For the entrainment events only trajectories that start with vorticity below the threshold are selected, and the first time at which the enstrophy threshold is overcome is identified as the time of interface crossing t_I. Conversely for extrainment events only trajectories that end with vorticity below the threshold are selected, and the last time at which the enstrophy decays below the threshold is tagged as the time of interface crossing t_I. In both cases, samples that cross the threshold before the steady state phase is settled are eliminated from the data set. Then, a new time reference system is defined as $t - t_I$ and conditional statistics are obtained by ensemble averaging all the samples at the same distance in time from the interface.

Bibliography

[1] BAGHERI, F. ; MITRA, D. ; P., Prasad ; BRANDT, L.: Statistics of polymer extensions in turbulent channel flow. In: *Physical Review E* 86 (2012), Nr. 5, S. 056314

[2] BALCI, N. ; BECCA, T. ; RENARDY, M. ; DOERING, C. R.: Symmetric factorization of the conformation tensor in viscoelastic fluid models. In: *Journal of Non-Newtonian Fluid Mechanics* 166 (2011), Nr. 11, S. 546–553

[3] BALKOVSKY, E. ; FOUXON, A. ; LEBEDEV, V.: Turbulence of polymer solutions. In: *Physical Review E* 64 (2001), Nr. 5, S. 056301

[4] BATCHELOR, G. K.: *The theory of homogeneous turbulence*. Cambridge university press, 1953

[5] BIFERALE, L. ; MUSACCHIO, S. ; TOSCHI, F.: Inverse energy cascade in three-dimensional isotropic turbulence. In: *Physical review letters* 108 (2012), Nr. 16, S. 164501

[6] BIRD, R. B. ; A., Robert C. ; HASSAGER, O. ; CURTISS, C. F.: *Dynamics of polymeric liquids*. Bd. 1. Wiley New York, 1977

[7] BISSET, D. K. ; HUNT, J. C. R. ; ROGERS, M. M.: The turbulent/non-turbulent interface bounding a far wake. In: *Journal of Fluid Mechanics* 451 (2002), Jan, S. 383–410

[8] BORRELL, G. ; JIMÉNEZ, J.: Properties of the turbulent/non-turbulent interface in boundary layers. In: *Journal of Fluid Mechanics* 801 (2016), Aug, S. 554–596

[9] BRADSHAW, P.: The understanding and prediction of turbulent flow. In: *The Aeronautical Journal (1968)* 76 (1972), Nr. 739, S. 403–418

[10] CASCIOLA, C. M. ; DE ANGELIS, E.: Energy transfer in turbulent polymer solutions. In: *Journal of Fluid Mechanics* 581 (2007), may, S. 419

[11] CHAUHAN, K. ; PHILIP, J. ; DE SILVA, C. M. ; HUTCHINS, N. ; MARUSIC, I.: The turbulent/non-turbulent interface and entrainment in a boundary layer. In: *Journal of Fluid Mechanics* 742 (2014), 3, S. 119–151

[12] CHEN, X. ; MARSCHALL, H. ; SCHÄFER, M. ; BOTHE, D.: A comparison of stabilisation approaches for finite-volume simulation of viscoelastic fluid flow. In: *International Journal of Computational Fluid Dynamics* 27 (2013), Nr. 6-7, S. 229–250

[13] CHENG, N. ; LAW, A. W.: Measurements of turbulence generated by oscillating grid. In: *Journal of Hydraulic Engineering* 127 (2001), Nr. 3, S. 201–208

[14] CIMARELLI, A. ; COCCONI, G. ; FROHNAPFEL, B. ; DE ANGELIS, E.: Spectral enstrophy budget in a shear-less flow with turbulent/non-turbulent interface. In: *Physics of Fluids* 27 (2015), Nr. 12

[15] CLASEN, C. ; PLOG, J. P. ; KULICKE, W. M. ; OWENS, M. ; MACOSKO, C. ; SCRIVEN, L. E. ; VERANI, M. ; MCKINLEY, G. H.: How dilute are dilute solutions in extensional flows? In: *Journal of Rheology* 50 (2006), Nr. 6, S. 849–881

[16] COCCONI, G. ; DE ANGELIS, E. ; FROHNAPFEL, B. ; BAEVSKY, M. ; LIBERZON, A.: Small scale dynamics of a shearless turbulent/non-turbulent interface in dilute polymer solutions. In: *Physics of Fluids* 29 (2017), Nr. 7

[17] CORRSIN, S. ; KISTLER, A. L.: The free-stream boundaries of turbulent flows. In: *NACA TN-3133, TR-1244* (1955)

[18] CRAWFORD, A. M. ; MORDANT, N. ; XU, H. ; BODENSCHATZ, E.: Fluid acceleration in the bulk of turbulent dilute polymer solutions. In: *New Journal of Physics* 10 (2008), Nr. 12, S. 123015

[19] DA SILVA, C. B. ; HUNT, J. C. R. ; EAMES, I. ; WESTERWEEL, J.: Interfacial Layers Between Regions of Different Turbulence Intensity. In: *Annu. Rev. Fluid Mech.* 46 (2014), S. 567–590

[20] DE ANGELIS, E. ; CASCIOLA, C. M. ; BENZI, R. ; PIVA, R.: Homogeneous isotropic turbulence in dilute polymers: scale by scale budget. In: *arXiv preprint nlin/0208016* (2002)

[21] DE ANGELIS, E. ; CASCIOLA, C. M. ; BENZI, R. ; PIVA, R.: Homogeneous isotropic turbulence in dilute polymers. In: *Journal of Fluid Mechanics* 531 (2005), 5, S. 1–10

[22] De Silva, I. P. D. ; Fernando, H. J. S.: Some aspects of mixing in a stratified turbulent patch. In: *Journal of Fluid Mechanics* 240 (1992), S. 601–625

[23] Dickinson, S. C. ; Long, R. R.: Laboratory study of the growth of a turbulent layer of fluid. In: *The Physics of Fluids* 21 (1978), Nr. 10, S. 1698–1701

[24] Dickinson, S. C. ; Long, R. R.: Oscillating-grid turbulence including effects of rotation. In: *Journal of Fluid Mechanics* 126 (1983), S. 315–333

[25] Dimitropoulos, C. D. ; Sureshkumar, R. ; Beris, A. N. ; Handler, R. A.: Budgets of Reynolds stress, kinetic energy and streamwise enstrophy in viscoelastic turbulent channel flow. In: *Physics of Fluids* 13 (2001), April, S. 1016–1027

[26] Doi, M. ; Edwards, S. F.: *The Theory of Polymer Dynamics*. Oxford science publications, 1986

[27] Draad, A. A. ; Kuiken, G.D.C. ; Nieuwstadt, F.T.M.: Laminar–turbulent transition in pipe flow for Newtonian and non-Newtonian fluids. In: *Journal of Fluid Mechanics* 377 (1998), S. 267–312

[28] Dubief, Y. ; Lele, S.K.: Direct numerical simulation of polymer flow. In: *Center for Turbulence Research: Annual Research Briefs* (2001), S. 197–208

[29] Dubief, Y. ; White, C. M. ; Terrapon, V. E. ; Sh., Eric S. G. ; Moin, P. ; Lele, S. K.: On the coherent drag-reducing and turbulence-enhancing behaviour of polymers in wall flows. In: *Journal of Fluid Mechanics* 514 (2004), 9, S. 271–280. – ISSN 1469–7645

[30] Elsinga, G.E. ; Marusic, I.: Universal aspects of small-scale motions in turbulence. In: *Journal of Fluid Mechanics* 662 (2010), S. 514–539

[31] Fattal, R. ; Kupferman, R.: Constitutive laws for the matrix-logarithm of the conformation tensor. In: *Journal of Non-Newtonian Fluid Mechanics* 123 (2004), Nr. 2, S. 281–285

[32] Fetters, L.J. ; Lohse, D. J. ; Colby, R.H.: Chain Dimensions and Entanglement Spacings. In: James, E.M. (Hrsg.): *Physical Properties of Polymers Handbook*. Springer, 2006, Kapitel 25, S. 445–452

[33] Fore, R. S. ; Szwalek, J. ; Sirviente, A.I.: The effects of polymer solution preparation and injection on drag reduction. In: *Journal of fluids engineering* 127 (2005), Nr. 3, S. 536–549

[34] Frings, B.: Heterogeneous drag reduction in turbulent pipe flows using various injection techniques. In: *Rheologica acta* 27 (1988), Nr. 1, S. 92–110

[35] Galanti, B. ; Tsinober, A.: Self-amplification of the field of velocity derivatives in quasi-isotropic turbulence. In: *Physics of Fluids* 12 (2000), Nr. 12, S. 3097–3099

[36] Gampert, M. ; Boschung, J. ; Hennig, F. ; Gauding, M. ; Peters, N.: The vorticity versus the scalar criterion for the detection of the turbulent/non-turbulent interface. In: *Journal of Fluid Mechanics* 750 (2014), S. 578–596

[37] Gampert, M. ; Kleinheinz, K. ; Peters, N. ; Pitsch, H.: Experimental and numerical study of the scalar turbulent/non-turbulent interface layer in a jet flow. In: *Flow, Turbulence and Combustion* 92 (2014), Nr. 1-2, S. 429–449

[38] George, W. K.: *Lectures in Turbulence for the 21st Century*. 2013

[39] Ghosh, I. ; McKinley, G. H. ; Brown, R. A. ; Armstrong, R. C.: Deficiencies of FENE dumbbell models in describing the rapid stretching of dilute polymer solutions. In: *Journal of Rheology* 45 (2001), Nr. 3, S. 721–758

[40] Graham, M. D.: Fluid dynamics of dissolved polymer molecules in confined geometries. In: *Annual Review of Fluid Mechanics* 43 (2011), S. 273–298

[41] Gyr, A. ; Bewersdorff, H. W.: *Drag reduction of turbulent flows by additives*. Bd. 32. Springer Science & Business Media, 1995

[42] Holzner, M. ; Liberzon, A. ; Guala, M. ; Tsinober, A. ; Kinzelbach, W.: Generalized detection of a turbulent front generated by an oscillating grid. In: *Experiments in Fluids* 41 (2006), Nr. 5, S. 711–719

[43] Holzner, M. ; Liberzon, A. ; Nikitin, N. ; Lüthi, B. ; Kinzelbach, A W .. W .and Tsinober T. W .and Tsinober: A Lagrangian investigation of the small-scale features of turbulent entrainment through particle tracking and direct numerical simulation. In: *Journal of Fluid Mechanics* 598 (2008), S. 465–475

[44] Holzner, M. ; Lüthi, B. ; Liberzon, A. ; Guala, M. ; Kinzelbach, W.: Entrainment Reduction and Additional Dissipation in Dilute Polymer Solutions. In: *Progress in Turbulence III*. Springer, 2009, S. 207–210

[45] Hopfinger, E.J. ; Toly, J. A.: Spatially decaying turbulence and its relation to mixing across density interfaces. In: *Journal of fluid mechanics* 78 (1976), Nr. 01, S. 155–175

[46] Jiménez, Javier: *Turbulence and vortex dynamics*. 2004

[47] Jovanovic, J. ; Pashtrapanska, M. ; Frohnapfel, B. ; Durst, F. ; Koskinen, J. ; Koskinen, K.: On the Mechanism Responsible for Turbulent Drag Reduction by Dilute Addition of High Polymers: Theory, Experiments, Simulations, and Predictions. In: *ASME. J. Fluids Eng.* 128 (2005), S. 118–130

[48] Kalashnikov, V.N.: Shear-rate dependent viscosity of dilute polymer solutions. In: *Journal of Rheology* 38 (1994), Nr. 5, S. 1385–1403

[49] Kalashnikov, V.N.: Degradation accompanying turbulent drag reduction by polymer additives. In: *Journal of Non-Newtonian Fluid Mechanics* 103 (2002), Nr. 2, S. 105–121

[50] Khare, R. ; Graham, M. D. ; de Pablo, J. J.: Cross-stream migration of flexible molecules in a nanochannel. In: *Physical review letters* 96 (2006), Nr. 22, S. 224505

[51] Kim, K. ; Li, C.F. ; Sureshkumar, R. ; Balachandar, S. ; Adrian, R. J.: Effects of polymer stresses on eddy structures in drag-reduced turbulent channel flow. In: *Journal of Fluid Mechanics* 584 (2007), S. 281

[52] Kinzel, M. ; Holzner, M. ; Lüthi, B. ; Tropea, C. ; Kinzelbach, W. ; Oberlack, M.: Scaling laws of turbulent diffusion-an experimental validation. In: *14th Int Symp on Applications of Laser Techniques to Fluid Mechanics Lisbon, Portugal*, 2008, S. 07–10

[53] Larson, R. G.: The rheology of dilute solutions of flexible polymers: Progress and problems. In: *Journal of Rheology* 49 (2005), Nr. 1, S. 1–70

[54] Liberzon, A. ; Guala, M. ; Lüthi, B. ; Kinzelbach, W. ; Tsinober, A.: Turbulence in dilute polymer solutions. In: *Physics of Fluids* 17 (2005), Nr. 3, S. 031707/1–4

[55] Liberzon, A. ; Holzner, M. ; Lüthi, B. ; Guala, M. ; Kinzelbach, W.: On turbulent entrainment and dissipation in dilute polymer solutions. In: *Physics of Fluids* 21 (2009), Nr. 3, S. 035107

[56] LIU, Y. ; JUN, Y. ; STEINBERG, V.: Concentration dependence of the longest relaxation times of dilute and semi-dilute polymer solutions. In: *Journal of Rheology* 53 (2009), Nr. 5, S. 1069–1085

[57] LONG, R. R.: A theory of mixing in a stably stratified fluid. In: *Journal of Fluid Mechanics* 84 (1978), Nr. 1, S. 113–124

[58] LÜTHI, B. ; TSINOBER, A. ; KINZELBACH, W.: Lagrangian measurement of vorticity dynamics in turbulent flow. In: *Journal of Fluid Mechanics* 528 (2005), April, S. 87–118

[59] MARK, J.E. ; FLORY, P.J.: The Configuration of the Polyoxyethylene Chain. In: *Journal of the American Chemical Society* 87 (1965), Nr. 7, S. 1415–1423

[60] MARTIEN, A. H. ; FATTAL, R. ; KUPFERMAN, R.: The log-conformation method in viscoelastic fluid flow simulations. In: *Workshop: Limits and Perspectives of Polymer Flow Modeling,July 8, 2005, Eindhoven*, 2005, S. 281–285

[61] MASOUDIAN, M. ; DA SILVA, C.B. ; PINHO, F.T.: Grid and subgrid-scale interactions in viscoelastic turbulent flow and implications for modelling. In: *Journal of Turbulence* 17 (2016), Nr. 6, S. 543–571

[62] MATHEW, J. ; BASU, A. J.: Some characteristics of entrainment at a cylindrical turbulence boundary. In: *Physics of Fluids* 14 (2002), Nr. 7, S. 2065–2072

[63] MCCOMB, W.D. ; RABIE, L.H.: Local drag reduction due to injection of polymer solutions into turbulent flow in a pipe. Part I: Dependence on local polymer concentration. In: *AIChE Journal* 28 (1982), Nr. 4, S. 547–557

[64] MCDOUGALL, T. J.: Measurements of turbulence in a zero-mean-shear mixed layer. In: *Journal of Fluid Mechanics* 94 (1979), Nr. 03, S. 409–431

[65] MIN, T. ; YUL Y.J.AND CHOI, H. ; JOSEPH, D. D.: Drag reduction by polymer additives in a turbulent channel flow. In: *Journal of Fluid Mechanics* 486 (2003), Jun, S. 213–238

[66] OBERLACK, M. ; GUENTHER, S.: Shear-free turbulent diffusion-classical and new scaling laws. In: *Fluid Dynamics Research* 33 (2003), Nr. 5, S. 453–476

[67] Owens, R. G. ; Phillips, T. N.: *Computational rheology*. World Scientific, 2002

[68] Pereira, A. S. ; Martins, R. S. ; Mompean, G. ; Thais, L. ; Thompson, R. L.: On the Extension of Polymer Molecules in Turbulent Viscoelastic Flows: Statistical and Tensor Investigation. In: *Progress in Wall Turbulence 2*. Springer, 2016, S. 171–180

[69] Perlekar, P. ; Mitra, D. ; Pandit, R.: Direct numerical simulations of statistically steady, homogeneous, isotropic fluid turbulence with polymer additives. In: *Physical Review E* 82 (2010), Nr. 6, S. 066313

[70] Prakash, J. R.: The kinetic theory of dilute solutions of flexible polymers: Hydrodynamic interaction. In: *Rheology Series* 8 (1999), S. 467–517

[71] Reynolds, W.C. ; Parekh, D.E. ; Juvet, P.J.D. ; Lee, M.J.D.: Bifurcating and blooming jets. In: *Annual review of fluid mechanics* 35 (2003), Nr. 1, S. 295–315

[72] Available under CC BY-SA 3.0.

[73] Silva, C. B. ; Pereira, J. C. F.: Invariants of the velocity-gradient, rate-of-strain, and rate-of-rotation tensors across the turbulent/nonturbulent interface in jets. In: *Physics of Fluids (1994-present)* 20 (2008), Nr. 5, S. 055101

[74] Silva, C. M. ; Philip, J. ; Chauhan, K. ; Meneveau, C. ; Marusic, I.: Multiscale geometry and scaling of the turbulent-nonturbulent interface in high Reynolds number boundary layers. In: *Physical review letters* 111 (2013), Nr. 4, S. 044501

[75] Sousa, P. C. ; Vega, E. J. ; Sousa, R. G. ; Montanero, J. M. ; Alves, M. A.: Measurement of relaxation times in extensional flow of weakly viscoelastic polymer solutions. In: *Rheologica Acta* 56 (2017), Nr. 1, S. 11–20

[76] Sreenivasan, K. R. ; Ramshankar, R. ; Meneveau, C.: Mixing, Entrainment and Fractal Dimensions of Surfaces in Turbulent Flows. In: *Proc. R. Soc. London, Ser. A* 421 (1989), Nr. 1860, S. 79–108

[77] Stoltz, C. ; de Pablo, J. J. ; Graham, M. D.: Concentration dependence of shear and extensional rheology of polymer solutions: Brownian dynamics simulations. In: *Journal of Rheology* 50 (2006), Nr. 2, S. 137–167

[78] Taveira, R. R. ; Silva, C. B.: Kinetic energy budgets near the turbulent/nonturbulent interface in jets. In: *Physics of Fluids* 25 (2013), Nr. 1, S. 015114

[79] Taveira, R. R. ; Silva, C. B.: Characteristics of the viscous superlayer in shear free turbulence and in planar turbulent jets. In: *Physics of Fluids* 26 (2014), Nr. 2

[80] Taylor, Z.J. ; Gurka, R. ; Kopp, G. A. ; Liberzon, A.: Long-duration time-resolved PIV to study unsteady aerodynamics. In: *IEEE Trans. Instr. Meas.* 59 (2010), Nr. 12, S. 3262–3269

[81] Tennekes, H. ; Lumley, J. L.: *A first course in turbulence*. MIT press, 1972

[82] Thompson, S.M. ; Turner, J.S.: Mixing across an interface due to turbulence generated by an oscillating grid. In: *Journal of Fluid Mechanics* 67 (1975), Nr. 02, S. 349–368

[83] Tordella, D. ; Iovieno, M.: Small-scale anisotropy in turbulent shearless mixing. In: *Physical review letters* 107 (2011), Nr. 19, S. 194501

[84] Townsend, A.A.: The mechanism of entrainment in free turbulent flows. In: *Journal of fluid mechanics* 26 (1966), Nr. 04, S. 689–715

[85] Tsinober, A.: Turbulent Drag Reduction versus Structure of Turbulence. In: Gyr, A. (Hrsg.) ; IUTAM Symposium (Veranst.): *Structure of Turbuelnce and Drag Reduction* IUTAM Symposium, Springer, 1990, S. 313–334

[86] Tsinober, A.: *An Informal Introduction to Turbulence*. Kluwer, 2001

[87] Tsinober, A.: *An informal conceptual introduction to turbulence*. Bd. 483. Springer, 2009

[88] Vaithianathan, T. ; Collins, L. R.: Numerical approach to simulating turbulent flow of a viscoelastic polymer solution. In: *Journal of Computational Physics* 187 (2003), Nr. 1, S. 1 – 21

[89] Valente, P. C. ; Silva, C. B. ; Pinho, F. T.: The effect of viscoelasticity on the turbulent kinetic energy cascade. In: *Journal of Fluid Mechanics* 760 (2014), 12, S. 39–62

[90] Vincenzi, Dario ; Perlekar, Prasad ; Biferale, Luca ; Toschi, Federico: Impact of the Peterlin approximation on polymer dynamics in turbulent flows. In: *Phys. Rev. E* 92 (2015), Nov, S. 053004

[91] VIRK, P.S.: Drag reduction in rough pipes. In: *Journal of fluid mechanics* 45 (1971), Nr. 02, S. 225–246

[92] VOTH, G. A. ; SOLDATI, A.: Anisotropic Particles in Turbulence. In: *Annual Review of Fluid Mechanics* 49 (2017), S. 249–276

[93] WARHOLIC, M.D. ; MASSAH, H. ; HANRATTY, T.J.: Influence of drag-reducing polymers on turbulence: effects of Reynolds number, concentration and mixing. In: *Experiments in fluids* 27 (1999), Nr. 5, S. 461–472

[94] WATANABE, T. ; GOTOH, T.: Coil-stretch transition in an ensemble of polymers in isotropic turbulence. In: *Physical Review E* 81 (2010), Nr. 6, S. 066301

[95] WATANABE, T. ; SAKAI, Y. ; NAGATA, K. ; ITO, Y. ; HAYASE, T.: Enstrophy and passive scalar transport near the turbulent/non-turbulent interface in a turbulent planar jet flow. In: *Physics of Fluids* 26 (2014), Nr. 10, S. 105103

[96] WATANABE, T. ; SILVA, C. B. ; SAKAI, Y. ; NAGATA, K. ; HAYASE, T.: Lagrangian properties of the entrainment across turbulent/non-turbulent interface layers. In: *Physics of Fluids* 28 (2016), Nr. 3, S. 031701

[97] WESTERWEEL, J. ; FUKUSHIMA, C. ; PEDERSEN, J. M. ; HUNT, J.C.R.: Momentum and scalar transport at the turbulent/non-turbulent interface of a jet. In: *Journal of Fluid Mechanics* 631 (2009), S. 199–230

[98] WOLF, M. ; HOLZNER, M. ; LÜTHI, B. ; KRUG, D. ; KINZELBACH, W. ; TSINOBER, A.: Effects of mean shear on the local turbulent entrainment process. In: *Journal of Fluid Mechanics* 731 (2013), S. 95–116

[99] XI, L. ; BAI, X.: Marginal turbulent state of viscoelastic fluids: A polymer drag reduction perspective. In: *Physical Review E* 93 (2016), Nr. 4, S. 043118

[100] YAN, J. ; CHENG, N.S. ; TANG, H.W. ; TAN, S. K.: Oscillating-grid turbulence and its applications: a review/Turbulence de grille oscillante et ses applications: une revue. In: *Journal of Hydraulic Research* 45 (2007), Nr. 1, S. 26–32

[101] ZHOU, Q. ; AKHAVAN, R.: A comparison of FENE and FENE-P dumbbell and chain models in turbulent flow. In: *Journal of Non-Newtonian Fluid Mechanics* 109 (2003), 2, Nr. 2-3, S. 115–155

Conference Contributions and Publications

The research work presented in this thesis resulted in the following conference contributions and publications:

COCCONI, G. ; CIMARELLI, A. ; FROHNAPFEL, B. ; DE ANGELIS, E.: A numerical study of the shear-less turbulent/non-turbulent interface. In: *Progress in Turbulence VI: Proceedings of the iTi Conference on Turbulence 2014*, Bertinoro, Italy.

COCCONI, G. ; FROHNAPFEL, B. ; DE ANGELIS, E. ; BAEVSKY , M. ; LIBERZON, A: Experimental and numerical investigation of turbulent entrainment in dilute polymer solutions. In: *15th European Turbulence Conference ETC 2015*, Delft, Netherlands

CIMARELLI, A. ; COCCONI, G. ; FROHNAPFEL, B. ; DE ANGELIS, E.: Spectral enstrophy budget in a shear-less flow with turbulent/non-turbulent interface. In: *Physics of Fluids* 27 (2015), Nr. 12

COCCONI, G. ; DE ANGELIS, E. ; FROHNAPFEL, B. ; BAEVSKY , M. ; LIBERZON, A.: Small scale dynamics of a shearless turbulent/non-turbulent interface in dilute polymer solutions. In: *Physics of Fluids* 29 (2017), Nr. 7

Schriftenreihe des Instituts für Strömungsmechanik

Karlsruher Institut für Technologie (KIT) | ISSN 2199-8868

Band 1 **Gertraud Maria Daschiel**
Strategies to reduce friction losses and their implications for the energy efficient design of internal flow domains.
ISBN 978-3-7315-0291-3

Band 2 **Andreas Güttler**
High accuracy determination of skin friction differences in an air channel flow based on pressure drop measurements.
ISBN 978-3-7315-0502-0

Band 3 **Anna Christina Daub**
Numerical Haemodynamics in the Human Heart.
ISBN 978-3-7315-0757-4

Band 4 **Alexander Stroh**
Reactive Control of Turbulent Wall-Bounded Flows for Skin Friction Drag Reduction.
ISBN 978-3-7315-0766-6

Band 5 **Giacomo Cocconi**
Numerical Study on the Propagation of Turbulent Fronts in Dilute Polymer Solutions.
ISBN 978-3-7315-0829-8

Die Bände sind unter www.ksp.kit.edu als PDF frei verfügbar oder als Druckausgabe bestellbar.